AF757707

# ISLAND TOURISM: SUSTAINABLE PERSPECTIVES

# Island Tourism: Sustainable Perspectives

*Editor*

**Regina Belo**

scitus academics

**Island Tourism: Sustainable Perspectives**

Edited by **Regina Belo**

ISBN: 978-1-68117-548-5
Library of Congress Control Number: 2016942952

© 2017 by
SCITUS Academics LLC,
www.scitusacademics.com
Box No. 4766, 616 Corporate Way,
Suite 2, Valley Cottage,
NY 10989

This book contains information obtained from highly regarded resources. Copyright for individual articles remains with the authors as indicated. All chapters are distributed under the terms of the Creative Commons Attribution License, which permits unrestricted use, distribution, and reproduction in any medium, provided the original author and source are credited.

**Notice**

Reasonable efforts have been made to publish reliable data and views articulated in the chapters are those of the individual contributors, and not necessarily those of the editors or publishers. Editors or publishers are not responsible for the accuracy of the information in the published chapters or consequences of their use. The publisher believes no responsibility for any damage or grievance to the persons or property arising out of the use of any materials, instructions, methods or thoughts in the book. The editors and the publisher have attempted to trace the copyright holders of all material reproduced in this publication and apologize to copyright holders if permission has not been obtained. If any copyright holder has not been acknowledged, please write to us so we may rectify.

Printed in United States of America on Acid Free Paper

# PREFACE

The concept of island tourism has attracted tourism researchers for ages who tried to examine islands' characteristics, focusing on development within an economic and social context, sustainability and effective management. Especially, in the last decades, with the developments in the infrastructure and the introduction of new air and sea connections, islands became easier accessible and more favorable by tourists. In fact, many of the world's small islands depend on tourism. So the relationship of tourism and island development is very important. It can be examined through various perspectives that reflect the complexity of interaction. For small island economies such as Okinawa, Hawaii, Guam and Saipan, tourism has been the fastest growing and most important industry accounting for 20-70% of their external receipts. These island economies have comparative advantages in tourism because of their small market sizes, rich marine resources, tropical or sub-tropical climate, rich and unique cultural heritages, hospitality-oriented peoples, service-intensive nature of tourism. Island tourism, however, heavily depends on islands' fragile and extremely limited natural as well as cultural resources. Islands' over-expanded tourism industry has created various socio-economic-ecological issues such as cultural friction, water shortage, environmental degradation, food insecurity, imported inflation and family problem on the life of islanders. Island tourism is not a recent phenomenon; the Romans used the Isle of Capri as a holiday destination two thousand years ago.

In order to understand the process of island tourism development, response to change and challenges and their journey to sustainability, *Island Tourism: Sustainable Perspectives* provides insights and instruction on topics including social, cultural, environmental and sustainable tourism. The present book examines islands and island tourism in relation to their specific characteristics as well as the impacts that tourism development has on islands. In addition, the development of tourism has positive and negative economic, socio-cultural and environmental impacts, that in the case of islands become more significant, since tourism is often more important for small islands than for mainland destinations.

# TABLE OF CONTENTS

1

# SUSTAINABLE TOURISM ON SEMAU ISLAND: READY OR NOT?

**Rode Ayu**

Sustainable Tourism & Research Development,
Dwi Nayika Solusindo, Indonesia

## ABSTRACT

While community development is globalized highly through small island opportunities worldwide, Semau Island is a natural Island that has great potentials for the development of tourism. Related to sustainability, moving from those early studies about sustainable tourism, it can be stated that, sustainable tourism seeks to minimize negative impacts on the local culture and natural environment while generating benefits for local residents. The community and visitor attraction is closely related. It means that tourism is ready to be developed and sustained on the Island. Besides, the methods provide primary and discussion data about whether or not there is a relation between tourism development and its impacts on local community readiness in a wide range of projects and sectors. The report examines the readiness responses from visitor attractions, community response and linkage on tourism to create number of visitors through their satisfaction value for sustainable tourism in Semau Island.

## KEYWORDS

Community participation; Readiness; Sustainable tourism; Tourism development; Visitor attraction

# 1. INTRODUCTION

Small islands have great potentials for the development of resourcebased industries, such as fisheries, manufacturing, and tourism. These islands will provide productive natural resources [1]. Small islands are potential resources in having coral reefs, sea-grass, mangrove, fishery resources and agriculture land areas. These resources would be tourism opportunity as well as visitor attraction to tourism development for current condition in Semau Island. Practically, the current condition must be managed well to be sustainable.

Tourism development depends heavily on environmental conditions, whether natural or man-made, for its market, its sustainability [2] and must be carried out in a way that is compatible with the principles of sustainable tourism development. McIntyre [3] defined sustainable tourism as an alternative tourism form that improves the quality of life of the host community, provides a high quality experience for visitors, and maintains the quality of the environment on which both the host community and visitors depend. Moving from those early studies about sustainable tourism, it can be stated that sustainable tourism seeks to minimize negative impacts on the local culture and natural environment while generating benefits for local residents. While sustainability has many definitions, it is fundamental to synchronize about good collaborative planning and about not spending more capital than possess – economic capital, environmental capital, or socio-cultural capital.

The problem of this research pointed to the Semau Island synchronization within readiness in planning and development of sustainable tourism. The key factor in achieving participatory planning is community 'readiness': i.e., Readiness with respect to tourism knowledge, resources, and commitment" [4] should be well managed. The Island is consists of host community, tourism attractions, environmental and sustainability that must be ready to be developed through tourism development management and practice ready can be described as a behavior. The research is concerning community respect related to readiness linkages on tourism and visitor attraction towards sustainable tourism. When the community and visitor attraction is closely linked, tourism on Semau Island is ready to be developed. The research methods provide primary data and discussion data about the relation between tourism development and its impacts on local community readiness or not in a wide range of projects and sectors. The analyses of the results gathered provide a valuable tool in order to establish some general conclusions

## 2. READINESS ON SUSTAINABLE TOURISM

Readiness in Change Management has been defined as the cognitive precursor to the behaviors of either resistance to, or support for, a change effort [5]. Furthermore, readiness is reflected in organizational members' beliefs, attitudes, and intentions regarding the extent to which changes are needed and the organization's capacity to successfully make those changes. In other words, community 'readiness': i.e., Readiness with respect to tourism knowledge, resources, and commitment should be well managed. The literatures identify that some level of readiness is necessary, it remains to determine, how and when community or organizational should be deemed ready to participate in the planning process to successfully make those changes. Knowledge as results from processing information in ways that are actionable [6] that used to a guideline to community for tourism information and development. Knowledge is contextual, relevant, actionable and can be used to solve problem has been defined in a pragmatic way. Knowledge can be classified as

(a) Declarative knowledge and

(b) Procedural knowledge [7]

Thus, declarative knowledge is factual knowledge of the industry such as data about tourist markets, environment, tourist behavior, competition and details of a destination and pprocedural knowledge includes the methodology used such as tourism planning models, communication tools, forecasting and quantitative methods.

Knowledge was different from information. Information relates to data, while knowledge involves a wider process that involves cognitive structures that assimilate information and put it into a broader context, thereby allowing actions to be undertaken on that basis. Information exists in dependently of the receiver and transmitter [8]. Knowledge is information that has been translated so that humans understand it. Knowledge cannot be said to 'flow' but can be said to be 'shared' or 'transferred'. Related to tourism products, knowledge with cultural and unique attractions including festivals and events, nightlife and performance facilities is needed so that tourists can see and experience the destination, including restaurants with diverse menus in different languages.

Tourism resources, the interpretation of the resource should, however, remain the key feature of the attraction visit, with increasing focus on catering for broader markets or more specific niche markets. Stage in classification usually focuses on the nature of the resource itself, be it natural or built. An accurate and honest assessment of the community's resources and possible tourist attractions addressed, both human and physical as one of the tourism Development assessment. Main reasons for this categorization result from the different approaches required for their management, with natural sites usually requiring fewer staff, incurring lower fixed costs and having a more open attitude towards access than in the case of built properties. Attractions are the raison d'être for tourism; they generate the visit, give rise to excursion circuits and create an industry of their own . The Island, a sample of destination product must be tries to be sucessful or fail tourist attractions and local destination management. An example of the island has been changed to create their own industry, Bali Island was well managed by Bali Tourism Board.

A community may be regarded as a repository of tourist resources by way of its bio centrality and ethno centrality [9]. The natural resources of a community (both physical and human) are often the raison d'être for the industry. Thus, a community offers its nature, culture, society and even its economy as commodities for tourism and to tourists. As a commodity, the community's intensive interaction with visitors is of utmost importance in the long-term sustainability of the industry since it is the culture and hospitality, along with the natural attraction, that create the image and experiences which attract visitors [10]. None the less, considerable development activity and income generated by tourism have not always been compatible with a location's social and economic objectives and can threaten the community's integrity.

Transformation in geography and typology of visitor attractions as tourism resources more generally, the visitor attractions sector arguably needs to adopt a more strategic approach to managing it activities to create quality relationship value for futures in local destination management. Commitment, generally described as a desire to maintain a valued relationship is recognized as one of the building blocks of relationship marketing among physical ttourism attraction and ccommunity. Related to community, Lauren Schlep Consulting [11] stated as a coordinated community commitment to plan and execute an effective strategy in terms of product and people to attract and satisfy visitors. When marketing approach tries to create quality relationship value with their effective strategy, community as host and resources must be agreed to have interrelationship to attract and satisfy visitors.

Three hypotheses were developed related to the model:

1. H1: declarative knowledge has a positive influence on community responses.
2. H2: natural resources have a positive influence on community responces toward visitor satisfaction on Semau Island.
3. H3: procedural knowledge has a positive influence on planning of tourism development. Hhypotheses above predicting community readiness and it components in tourism development process to increase tourists number will be portrait on the readiness framework below (Figure 1).

# 3. METHODOLOGY

The causal and explanatory research design are used to develop research model aims to test research hypotheses have been determined and expected. "Nor does the ability to predict accurately demonstrate anything about causality" [12]. Visual observation and a cross-sectional analysis or sample survey with a self-administered questionnaire were used to collect data. All closed-ended questions in the questionnaire were developed by reviewing relevant literature. The questionnaire was comprised of three sections. The first section was designed for demographic information divided into three constructs, they are: sex (male and female), education (high school, bachelor and master) and age (youth, adult, citizen). The second section was designed to measure three constructs of readiness, including six dimensions of knowledge (declarative and procedural), resources (physical and natural attraction) and community commitment (planning and satisfaction). 17 items adapted from prior empirical studies were measured using a five-point Likert-type scale (1: totally disagree to 5: totally agree). The last section of the questionnaire was designed to capture respondents' demographic characteristics (Table 1).

Before distributing the survey questionnaire, the pilot test was conveniently circulated to graduate students in a southern university to identify the appropriateness and wording of the questions, sequence of the questionnaire, and analysis procedure. A total of 29 responses were analyzed for crosschecking the reliability of the measures. Cronbach's alpha of knowledge variable (α = 0.993), resources variable (α = 0.813) and commitment variable (α = 0.851) were all higher than the lower limit of 0.70 suggested by Hair. The result above clarified that all instruments had internal consistency or reliability and validity with correlation analysis (**. Correlation is significant at the 0.01 level 2-tailed).

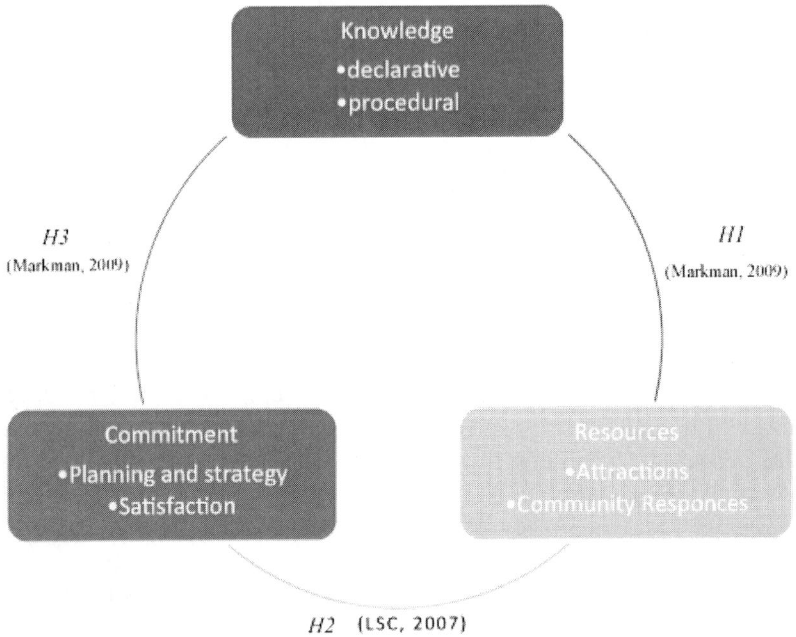


**Figure 1.**Sustainable tourism readiness framework

Instruments was distributed in community along six months from November 2012 to April 2013 to collected the data. A non-probability sample with systematic sampling was used to draw samples. A total of 430 questionnaires were distributed to selected subjects directly, but only 405 successfully delivered. Of these, 344 participants responded to the survey. After removing the cases with excessive levels of missing data, 310 responses remained in the analysis. Then, the SEM Analysis was developed to find the model and value of tourism readiness on Semau Island.

**Table 1.** Items of instruments

| No | Descriptions | Initial |
|---|---|---|
| 1 | I have good benefits from tourism | DK1 |
| 2 | My life change because tourism be a part of my activities | DK2 |
| 3 | Create job for local residents | DK3 |
| 4 | Employs local youth | DK4 |
| 5 | Raises prices for goods | PK1 |
| 6 | Helps community obtain services | PK2 |
| 7 | Causes rise in crime rates | PK3 |
| 8 | Harms moral standards | PK4 |
| 9 | Disturb local activities | PR1 |
| 10 | Stops locals from beach disturbsion | PR2 |
| 11 | Helps stimulate local culture and crafts | CR1 |
| 12 | Use natural resources needed by local residents | CR2 |
| 13 | The community has control over tourism | SC1 |
| 14 | The money spent by tourists remains in my community | SC2 |
| 15 | Local residents have easy access to the area which tourist use | SC3 |
| 16 | Satisfactory opinion | SAC1 |
| 17 | Want more tourism futures | SAC2 |

# 4. RESULTS AND DISCUSSION

## 4.1 Semau island profiles

Administratively, Semau Island is one of small islands and part of Kupang Regency, East Nusa Tenggara province located in the north of Timor Island. More than 10.000 residents live there. The island area is about 246.66 km2 and the total of land use is 97 percent from the total area. Moreover, Semau Island is divided into two districts, such as: Semau district and the capital city is Uitao, with eight villages and 35 sub villages, and South Semau district and the capital city is Akle, with six villages and 27 sub-villages.

Topographically, 60 percent of Semau Island has sloping surface with 3-20 meters height above sea level. As for the village area Uitao, Otan and Huilelot villages have a surface of highland with an altitude up to 243 meters sea level above. Thus, the potential of the agricultural sector, farming and fishing are the main occupation on the Semau Island to develop the product. Those activities are supported by the island and ecosystems, local commodity which has many variance of strategic location on the island.

Geographically, the coordinate of Semau Island is located at 10° 10° 14' 0" LS 123° 23' 30" BT. Semau Island is a flat area with low rain debris and has a tropical climate. Rainy or Wet season will be started from December to February and summer season up to 37 degrees started from March to October. It has many potencies of water source actually. Not to mention, the beach condition with various sea biotic and white sandy beach will show a beautiful scene for tourists (Figure 2).

Indigenous culture called as Helong society had unique ethnic and cultural heritage can be developed as community and cultural tourism and alternative in modern mass tourism. Thus, these heritage such as: Lingae dance, corn ritual and harvest moment needs to be conservated. Unfortunately, since the recent traditional culture was affected by Christian since 1980s, the indigenous community of Semau or Helong society destroyed the traditional culture and heritage from their ancestors.

## 4.2 Respondent profiles

The sample of 310 respondents comprised males (41.6%) and females (58.4%). The majority of respondents were youth (58%). About 34% of the respondents were adult and the rest of respondents were the silent generation or citizen. Over 48.1% of the participants reported having completed high school, with 35.2% having bachelor degrees and 16.8% having master degrees.

## 4.3 Testing hypothesis model

The proposed conceptual framework was tested by AMOS structural equation modeling (SEM) with a two-step approach recommended by Anderson and Gerbing. With this approach, a confirmatory factor analysis (CFA) was first performed to establish the acceptable levels of goodness-of-fit with the measurement model, and then SEM was performed to test the hypothesized model. The default model of readiness framework is significant because P value=0.00 with $^{**}P \leq 0.05$).

Undertaking CFA, a total of twenty-four measured variables were constrained into three hypothetical constructs as follows: Knowledge variable (two items), Resources variable (two items) and Commitment variable (two items). The goodness-of-fit statistics (X2/df ratio=2.75, RMSEA=0.250 [CI for RMSEA=0.11, 0.398], PCLOSE=0.017, CFI=0.973, GFI=0.857, AGFI=0.501, TLI=0.934, SRMR=0.032, NFI=0.960) fell within the acceptable ranges, signifying that the overall measurement model provided an adequate fit to the data. The reliability and validity of the constructs were also tested. The analysis revealed that all seven constructs had Cronbach's alpha higher than the lower limit of 0.70 [13]. The test revealed that all value estimates were greater than the corresponding inter constructs squared correlation, supporting discriminant validity of the measurement model (Table 2).

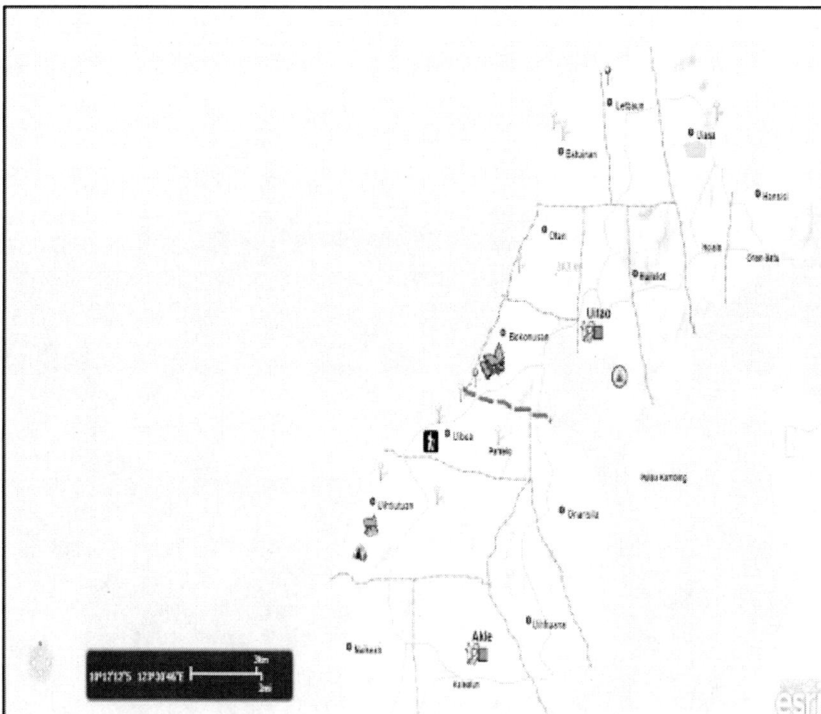


**Figure 2.** Semau Island Map, East Nusa Tenggara, Indonesia. The map was created data and content from GIS map-Ina; ESRI produced, 2013.

**Table 2.** Sample Correlations (Group number 1).

| | Declarative | Procedural | Physical | Community | Planning | Satisfaction |
|---|---|---|---|---|---|---|
| Declarative | 1,000 | | | | | |
| Procedural | ,042 | 1,000 | | | | |
| Physical | ,261 | ,923 | 1,000 | | | |
| Community | ,056 | ,953 | ,970 | 1,000 | | |
| Planning | ,059 | ,953 | ,970 | ,999 | 1,000 | |
| Satisfaction | ,058 | ,947 | ,966 | ,995 | ,994 | 1,000 |

Condition number = 9227,178
Eigenvalues4,881 1,024,075, 013,006,001

**Table 3.**Testing Hypotheses Model.

| Relationships | Estimate | S.E. | C.R. | P | Correlations | Hypotheses |
|---|---|---|---|---|---|---|
| Knowledge<-->Resources | 0.541 | 0.044 | 12.401 | *** | 0.261 | Accepted |
| Knowledge<-->Commitment | 0.516 | 0.043 | 12.105 | *** | 0.966 | Accepted |
| Resources<-->Commitment | 0.516 | 0.043 | 12.130 | *** | 0.953 | Accepted |

After the overall measurement model was tested by CFA and shown to have adequate fit, as well as construct validity, SEM was followed to test the hypothesized structural relationships among constructs presented. In addition the result of SEM revealed that all six structural path estimates were statistically significant in the expected direction.

Sample correlations and Testing hypotheses model resulted that H1 is accepted that declarative knowledge has a positive correlation value 0.261 on physical resources and knowledge has influence value 0.541 on community responses (Table 3). It can be argued that improving declarative knowledge of tourism is the central prerequisite for enhancing community participation tourism development [14] in Semau Island. Eventhrough low correlation value has been predicted, declarative knowledge related to tourist markets, environment, tourist behavior, competition and details of a destination as important thing for community must be well introducted. A tourism destination products and services must constantly evolve and innovate to ensure that the demands and needs of visitor markets are consistently met. More than 10.000 residents have to know and responces about tourist and their behavior to attract more tourist and increase benefits from tourism activities.

Moreover, H2 is accepted that natural resources have a positive correlation value 0.966 on community responces toward visitor satisfaction on Semau Island and influence value 0.516 toward commitment. Natural attractions can be natural site and event or activities. Tourism activites as commodity on sites create community responces to participate as hosts. As commodity, the community's interaction is very important for long-term sustainability, along with the natural attraction, that create the image and experiences which attract visitors as satisfaction of host services. The assessment examines both existing attractions and attractions that have half potential for future development or sustainable tourism and ready in going to sustainable tourism development for better changes.

Thus, the last hypothesis is accepted that procedural knowledge has a positive correlation value 0.953 on planning of tourism development and influence value is 0.516 toward knowledge. Lauren Schlep Consulting stated that "a coordinated community commitment to plan and execute an effective strategy in terms of product and people to attract and satisfy visitors". Methodology and marketing approach as part of planning structure will develop to create tourism development on Semau Island. When marketing approach tries to create quality relationship value with their effective planning, community as host and resources must be agreed to have interrelationship to attract and satisfy tourists. It means that community on Semau Island is absolutely ready to go to sustainable tourism development for better changes trough planning process.

## 4.4 Linking knowledge, resources and community commitment towards sustainable tourism development

Tourism will be main concern if community and natural attractions syncronize within tourism activities. Community responces and participation in tourism development in poor rural areas has the potential capacity for increasing incomes and employment, developing skills and institutions, and thereby empowering local people [15]. Increase benefits (income generating, knowledge, etc.) from sustainable tourism resources for communities on Semau Island. Solutions to keep visitor attraction, host communities (local or indigenous) as a stakeholder to be sustainable or match with tourism development plan and strategy.

Brourke and Luloff said that community 'readiness': i.e., Readiness with respect to tourism knowledge, resources, and commitment must be developed. The literatures identify that some level of readiness is necessary, it remains to determine, how and when community or organizational should be deemed ready to participate in the planning process to successfully make those changes. According to the finding above, successfully planning for tourism development requires the active support and involvement of the local community. Community participation as one of tourism plan elements in tourism development [16,17]. Tourism development must be created and well managed to sustainability. It is important to recognize that host communities have a stake in ensuring that tourism does not compromise their quality of life. Then all of indicators of sustainable tourism and development planning process will ready to implement tourism planning into development to sustainability.

# 5. CONCLUSION

Tourism development must be carried out in a way that is compatible with the principles of sustainable tourism. Moving from those early studies about sustainable tourism, it can be stated that, sustainable tourism seeks to minimize negative impacts on the local culture and natural environment while generating benefits for local residents. Then, related to tourism readiness on Semau Island, it can be explained that they have good knowledge related to their opportunity to increase benefit in tourism industry, half Resources availability and high commitment related to satisfactory result. It means that Community is ready to increase Tourism Development on Semau Island. Some level of community readiness is necessary, it remains to determine, how and when a community should be deemed ready to participate in the planning till development process.

The main objective of the research stated clearly through the analysis that is linking community respect on commitment in tourism and visitor attraction towards sustainable tourism that will be seen from visitor satisfaction. When the community and visitor attraction is linked enough, it means that tourism is ready to be developed and sustainable in the Island. The opportunity can be described as recommendation for the government that visitor attraction must be developed well. It will be linked by lacking of support services and community participation in increasing number of visitor on Semau Island because of their readiness response for sustainable tourism.

# REFERENCES

1. Retraubun, ASW (2003)Prospects of small island development, Directorate General of Coastal Zone and SmallIslands of the Department of Marine and Fisheries, Jakarta.
2. Tisdell CA Tourism Economics, the Environment and Development. Analysis and policy, Cheltenham.
3. McIntyre G (1993) Sustainable TourismDevelopment: Guide for Local Planners. World Tourism Organization, Madrid.

4. Bourke L, Luloff AE (1996) Rural Tourism Development is Community in Southwest Rural Pennsylvania Ready to Participate, John Wiley & Sons, USA.

5. Armenakis, AA et al. (1993) Creating readiness for organizational change, Hum Relat., 46: 681–703.

6. Alavi M, Halley B (2009) Knowledge management systems: Implications and opportunities for data warehousing. Journal of Data Warehousing. 4: 2-6.

7. Markman AB(2009) Knowledge representation. Lawrence Erlbaum, London.

8. Chang YH, Chen MH (2004) Comparing approaches to systems of innovation: the knowledge perspectiveTechnol Soc.26: 17–37.

9. Murphy PE (1985) Tourism: A Community Approach. Routledge, London.

10. Haywood K.M (1988) Responsible and responsive tourism planning in the community. Tourism Manage.9: 105–118.

11. Lauren Schlep Consulting (2007).

12. Campbell A, Converse PE, Rodgers WL (1976) The Quality of American Life:Perceptions, Evaluations, and Satisfactions. Russell SageFoundation, New York.

13. Cronbach LJ (1951) Coefficient alpha and the internal structure of tests.Psychometrika.16:297-334.

14. Moscardo G(2008)Building community capacity for tourism development. CABI publisher, UK.

15. Ashley C, Garland E(1994) Promoting Community Based Tourism Development: Why, What and How?

16. Tosun C, Timothy DJ (2003) Tourism Growth, National Development and Regional Inequality in Turkey. Journal of Sustainable Tourism. 11: 133-161.

17. Timothy DJ, Tosun C (2003) Appropriate planning for tourism in destination communities:participation, incremental growth and collaboration. CAB International, UK.

2

# POLITICAL ECOLOGY, ISLAND TOURISM PLANNING,AND CLIMATE CHANGE ADAPTATION ON BORACAY, PHILIPPINES

**Virgilio Maguigad, David King, Alison Cottrell**

College of Marine and Environmental Sciences,
James Cook University, Australia

## ABSTRACT

This research paper presents a case study of the island of Boracay, Philippines, utilising a political ecology approach to climate change adaptation. The research finds that the island's political ecology, especially the relationships among stakeholders, is strained. This creates challenges for various urban planning processes that require good working relationships. Climate change is expected to highlight these divisions as interactions among stakeholders (fulfilling zoning ordinance obligations, climate change adaptation- compliant land use plans, *etc.*) are dependent on good stakeholder relations. Stakeholders realise that climate change is real and that sea level rise is already challenging existing zoning ordinances on urban beach development. However, this realisation must be integrated into political decision-making processes involving tourism stakeholders. The research also shows that the political ecology approach and methodology is applicable to studying the dynamics of climate change adaptation and tourism urbanisation on small islands.

## KEYWORDS

Climate change adaptation, island tourism planning, Philippines, political ecology, tourism urbanization

# 1. INTRODUCTION

The theoretical and methodological foundations for the research described in this paper are set out in detail in Maguigad, King, and Cottrell (2015). In this previous paper, we discussed how global climate change phenomena are experienced locally in diverse ways, despite being widely identified as a threat to coastal and island communities, including the Philippines. Similarly, we discussed how tourism has been identified as a vital economic activity in many developing economies, including the Philippines, despite also being identified as a contributor to climate change (Broderick, 2009; Dubois *et al.*, 2011). Climate change and tourism thus have a tense relationship in a country such as the Philippines, which is experiencing the economic benefits of strong tourism growth (DOT, 2012; Arnaldo, 2012) alongside a recognition that it is highly vulnerable to climate change (Harmeling, 2011) and that its urban populations are exceptionally at risk from rising sea levels (Dasgupta*et al.*, 2012; Nicholls *et al.*, 2007).

The present paper focuses on Boracay Island (Malay municipality, Aklan province) in the Philippines. Boracay is a major and well-documented tourism destination. Boracay has been an overwhelming success in terms of the tourism industry, and its visitor arrivals outnumber all other Philippine small island destinations according to Department of Tourism statistics. However, current development on the island, which has been subject to rapid tourism urbanisation, points to a stagnation stage in the Tourism Area Life Cycle (TALC) model, with massive intervention needed to avoid environmental, infrastructural, and visitor experience degradation (Smith *et al.*, Indeed, although Boracay is frequently portrayed as a tourism gem, it is cited in the literature as an example of tourism gone wrong, with a myriad of environmental and social problems (Carter, 2004; Ong *et al.*, 2011). As early as 1997, researchers sounded the alarm on the need to remain within physical, tourist, resident, transport, and governance carrying capacity thresholds if Boracay was to continue as a sustainable destination (Trousdale, 1997, 1999). The growth since the 1990s has so far resulted in a situation characterised by *ad hoc* coastal tourism development, emphasis on improvement and cleanliness of the 'visual environment' but lacking in social and cultural sustainability, and mutual adjustment among various stakeholders (Carter, 2004; Ong *et al.*, 2011). The island is deemed very important to the Philippines' tourism, and its challenges are currently co-managed by the national government through the Department of Tourism (DOT), Department of Environment & Natural Resources (DENR), and the local government of Malay, thereby going against the spirit of local governmentdevolution. Nevertheless, the Malay municipal planning and development officer Alma Belejerdo (2012) acknowledges the municipality's challenges and admits that the municipality needs all the help it can get from both the national government and the private sector.

**Figure 1.** A view of the main beach of Boracay during the southwest monsoon season (Author's Fieldwork Photos, August 2012).

On the basis of research undertaken on Boracay, this paper presents the intimate political ecology dynamics of an urbanised tourist island in the context of climate change adaptation. The adaptive capacity of the agents, systems, and institutions internal and external to the islands' tourism system with respect to climate change is dependent upon various factors, one of which is how these agents or actors perceive their roles and participation in the culture of tourism, climate change adaptation, and governance. Unravelling these factors and perceptions is best done by applying the theoretical and methodological political ecology approach set out in the previous paper (Maguigad, King, & Cottrell, 2015).

## 2. METHODS

Fieldwork was undertaken on Boracay during three distinct periods: August 2012 for preliminary preparatory fieldwork, including securing permissions from the Malay local government to undertake the research; July-August 2013 for the main fieldwork, including interviews and small group meetings; and April 2014, including final interviews and opportunities to react to preliminary results of the 2013 interviews, which were made available to selected key stakeholders. Human ethics and fieldwork methodology approvals were obtained from James Cook University prior to the commencement of the fieldwork.

A total of 20 interviews and small group meetings with a total of 27 participants were undertaken, with a majority of the interviews recorded using a digital voicerecorder. The interviews were conducted in a mixture Filipino (the national language)and English (the second official language). The researcher had previously prepared guide questions for the interviews and discussed the overall objectives of the PhDresearch with each of the respondents. If and when consent was secured, the interviewproceeded in a semi-structured process, though hewing closely to the previouslyprepared guides. This method of data collection tends to result in rich discussions anda wealth of non-quantifiable data that reveals a complex web of personal experiences,which would otherwise be lost in traditional methods such as questionnaire surveys. This method is highly suited to the political ecology framework.

The questioning technique and delivery of questions to the respondents was made in such a way as to avoid preconditioned, stock responses. Because we sought to uncover the nuances of the relationships between the various tourism actors on the island in the context of climate change adaptation, the questioning technique required a delicate consideration of local customs, trust between the researcher and respondents, and sometimes 'beating around the bush' and small talk as well as and indirect questioning in order to obtain answers. The semi-structured interviewtechnique allowed the researcher to ask unprompted and prompted questions. Unprompted questions usually inquired into the respondents' backgrounds, working and living conditions, and experiences on the island. Prompted questions would relate to the research's themes on climate change and tourism, which were explained before the start of the interviews. The nature of relationships among stakeholders would be called into question as these would emerge naturally from the follow-up questions fielded throughout the interview.

There were instances when some critical and sharp-thinking respondents asked the researcher why questions would revolve around certain topics, for instance, experiences concerning public consultations on tourism and land use planning, and how this relates to climate change (*e.g.* 'I thought you were asking about tourism and climate change, so why are you asking about the local government?'). These were isolated instances, however, and presented further opportunities to deepen the discussion with the respondents. While conducting the interviews and small group discussions, the responses would vary and could include long-winded answers that might or might not be related to the question asked. It is during these rambling moments that some other underlying beliefs, serendipitous, or volunteered privileged information might be shared, which would have been left uncollected by a less flexible questioning technique.

The interviews and small group meetings totalled 1114 minutes (18 hours and 34 minutes) and were translated and transcribed by the researcher in Nvivo, generating 187,158 words of codable transcript. The transcript was coded according to the emerging themes in each of the paragraphs of transcript, consistent with the ethnographic research framework. Aided by the use of Nvivo, all subsequent stakeholder analyses were then derived from these themes. A total of 103 themes were generated from the interview transcripts, and each of the respondents were grouped into five stakeholder nodes, namely: Visitors, Tourism Businesses, Government, Residents, and Non-Governmental Organisations. These groups correspond to the stakeholder groups that were identified as most likely to be involved in the tourism sector. Each of the 27 Boracay respondents who participated in the interviews and small group discussions was classified into the five groups, with the possibility that they could belong to more than one group. For example, a resident could also be classified as belonging to the government sector by virtue of working with any of the government offices. I have classified tourism workers who have worked on the island for more than five years as residents as it is deemed that they have acquired some perspective that would be comparable to that of local island residents.

Coding and theming as well as the resulting theme names were derived directly from the respondents' own words as translated into English. The 'Five Ways of Being' (Thompson *et al.*, 1990) in Cultural Theory (CT), using the pioneering studies by anthropologist Mary Douglas, argues that the extent of individuals' involvement in community life depends on the extent to which they are immersed in 'groups' as well as how far their lives are affected by external factors (termed 'grids') in CT(Forsyth, 2003). It is thus expected that although respondents may belong to differentstakeholder groups (*e.g.* resident and government, tourism sector and resident), their responses may vary depending on their affinity to these groups. During the interviews, some would express 'personal views' that would not reflect their groups' position, and these would be coded according to themes that have emerged.

Nvivo is a tool that has made qualitative research more systematic and taps into the technology to gather, record, transcribe, visualise, organise, and quantify qualitative data previously believed to be too unwieldy and 'messy' to handle using traditional methods (Welsh, 2002). Notwithstanding the debates and discussions that such an approach stirs in academic circles (Brown *et al.*, 1990), we believe that Nvivo enhances rather than interferes with the capabilities of any researcher undertaking this type of research.

We used Nvivo's Pearson's correlation functionality to measure similarity between pairs of nodes (the resulting themes) as well as the sources of the transcripts (stakeholder grouping). To form cluster diagrams, Nvivo first builds a table in which the 103 resulting nodes (themes) are paired with one another to calculate the degree to which they are related using Pearson product-moment correlation coefficient (PMCCC), more simply known as Pearson correlation coefficient. This is a measure of correlation or dependence between two variables, which is scaled from +1 (total positive correlation) to -1 (total negative correlation), with a value of zero indicating no correlation (NVIVO, 2014). The correlation analysis was carried out between two themes or nodes, between two stakeholder groups, and between a theme and a stakeholder group.

A summary of the clustering processes undertaken using Nvivo is provided in Table 1 below

The Pearson correlation coefficient computation was done using Nvivo with a total of 5779 pairings between and among 103 coded themes and 5 stakeholder classifications. The coefficients generated ranged from -0.013251 between the nodes/themes 'greenwashing' and 'emergency services' to a near-perfect correlation between the themes generated within the stakeholder groups 'Tourism Business' and 'Residents'. The sheer amount of correlation computations done with the aid Nvivo certainly adds to the richness of perspectives, which would never have come to light if researched using traditional means. This facility enabled the researcher to delve into the themes that help create a picture of the island's political ecology that could be relevant to the mainstreaming of climate change adaptation on tourist islands.

**Table 1.** Nvivo cluster analysis done via Pearson's correlation.

| Table rows | Clustered by | Table columns | Table cells |
|---|---|---|---|
| **Sources (5 stakeholder groups)** | Word similarity | Each different word that appears in the text of the sources | The number of times the column's word appears in the row's source |
| | Coding similarity | Each node that codes the sources' content | 1 if the column's node codes the row's source, 0 otherwise |
| **Nodes (103 themes)** | Word similarity | Each different word that appears in the text of the nodes | The number of times the column's word appears in the row's node |
| | Coding similarity | Each source coded by the row's node | 1 if the column's source is coded by the row's node, 0 otherwise |
| **Words (top 100 words in Word Frequency query results)** | N/A | Each source or node that in which the query searches | The number of times the row's word appears in the column's source or node |

## 3. STAKEHOLDERS' VIEWS AND LIVED EXPERIENCES IN THEIR URBANISING ISLAND ENVIRONMENT

Boracay is frequently mentioned by the national tourism organisation as the crown jewel of Philippine tourism, and one of the marketing foundations for the island is that it is dubbed as having the world's best beaches (ABS-CBN News, 2013; Ong *et al.*, 2011). This claim is at best valid for the majority of the interviewed stakeholders: "Because of the strong marketing being done for Boracay, it has become a tradition for some families or visitors to make it a tradition to visit the island" (TourismBiz_JCS, 2013).

There is often a sense of awe and excitement that prevails when first-time visitors alight on the island, and the experiences they have set them on course for repeated visits. One frequent visitor to the island enthuses:

I first came here together with my partner who has just come from Mexico, his sister who already visited the Maldives, and myself which at that time I have recently visited Bali, and when we arrived, we all fell in love with the island at that very moment. Boats then would bring passengers right in front of the boat stations, so the moment our feet touched the sand, we fell in love. The quality of sand is still the same after all these years (Visitor_ACG, 2014).

All stakeholders recognise Boracay's endowment of superfine white sand, and they acknowledge that it is the reason why visitors return to the island: "For first timers, visiting Boracay is really an exciting experience for them with the clean waters and the superfine sand as being commonly mentioned as the main attractions" (TourismBiz_ML, 2013). However, the awareness of the island's beachfronts and foreshores as an attraction and destination (P coefficient 0.812222) is also tempered by the stakeholders' comments that their beachfront is confronted by beach erosion (Pearson coefficient 0.75924). Stakeholders have noted:

There was a time at Station 1 where the roots of the trees near the beach were already showing, which points to beach erosion. I think it has been a problem for several years now. Boracay is always in the news for these types of issues (TourismBiz_LR, 2013).

There are subtle differences in the way beach erosion is viewed among stakeholders, probably due to the roles and the groups to which the respondents belong. For example, a provincial-level environment official views beach erosion as something that can be 'managed', reflecting the corresponding Cultural Theory representation of the 'managerial role' played by governments in managing environmental concerns (Forsyth, 2003) as they view nature as "perverse and tolerant." The provincial government official says:

Other environmental concerns of Boracay would be beach erosion, and there would be scouring in times of active wave activity like storm surges. The locals, however, say that it is only natural for sand to be eroded but then it will return during the *habagat* monsoon season. The provincial government is studying the possibility of beach nourishment activities as are being done in Florida, but it has never been done. There were concerns that the beach sand that will be used to replace eroded beachfronts will not be as white as the existing sand (Govt_JKA, 2013).

**Figure 2.** Eroded sections of White Beach on Boaracay. The high tide reaches up to the roots of the coconut trees. The visible concrete aggregates are remains of the banned semi-permanent concrete structures that have been illegally constructed by the resorts in violation of beach setback rules.

Long-time residents who are engaged in the tourist trade have also noted that beach erosion has become more severe and more noticeable. They are simultaneously critical of the comments by the Department of Environment & Natural Resources (DENR) offices regarding the reasons for erosion, saying that:

The Diniwid beach area is where the beach erosion has been severe, near Microtel after the grotto after passing Station 1. The DENR said that the bagging of sand and the illegal embankments that were constructed have caused more erosion as it impeded the flow of sand. However, it is counter-intuitive as the southwest monsoons have eroded much of the beach in that part of the area. The sloping beach has become a sudden drop where the water gets deep so close to shore (Resident&TourismBiz_JDT&JCT, 2013).

Tourism business operators are easy targets for the provincial government, which accuses most resort owners of violating environmental laws. That is, tourism urbanisation is blamed for exacerbating erosion. The provincial government thus places responsibility for uncontrolled growth near the high water mark of the beach on the Malay municipality, which has land use and tourism oversight of the island. One of the managers of a resort mentions that:

There has been a big discussion related to the 25+5 meter rule [easement] as the sea seems to be 'getting the sand', where the sand is really removed from the beachfront, especially in Diniwid. They say that because the resorts are putting sandbags, fences, or illegal structures that beach erosion has been more prevalent (TourismBiz_RA, 2013).

The acknowledgement of beach erosion and its link to sea level rise (refer to Table 3 for values) is not lost on various stakeholders, though the issue can be confused with extreme weather events like typhoons. One of the island residents, who is also a high-ranking municipal official, recounts:

There has been no change in typhoon patterns here in Boracay, and we still get typhoons. The last major typhoons we experienced were Typhoons Seniang and Frank, maybe in the last two to three years. We usually get sea level rises during major typhoons, especially along the shorelines. When we were kids, we usually go to the beach when the seas churn due to the coming storms (GOVT&Resident_FDLS, 2013).

However, the manner in which planning interventions have been interpreted by the various stakeholders vary, reflecting the biases inherent to their groups. The Aklan provincial government official, for example, relates:

After proclamation PD1064, the buffer zone of 25+5 meters must be maintained on the beaches. After that, the NAMRIA has designated markers for measuring setbacks. However, if they measure from the highest tide, nothing will be left to be built as we will have to demolish everything. This will even be compounded by sea level rise and storm surges. That is something the public cannot understand as they think that we are just making their lives difficult, but in fact it is only for their safety (Govt_JKA, 2013).

A regional government official representing the Housing and Land Use Regulatory Board(HLURB), the national agency that promulgates land use and planning guidelines for local governments, is even more critical of the exceptions accorded on Boracay:

According to the Water Code of the Philippines, there should be an easement of 40 meters from the highest tide of the shorelines, but they are insisting on the 25+5 meter rule. We are actually advocating not only land use but land and water use planning. The reason why their CLUP [comprehensive land use plan] could not pass the provincial land use council and the provincial council is precisely because the law says 40 meters' easement. One can go beyond the legal requirement but not below what the law prescribes (Govt_RL, 2014).

The open conflict between the various levels of government – the municipal government of Malay, the provincial government, and the various national government agencies – on the interpretation of setbacks has ramifications for physical planning. It also has an impact on the more contentious issue of demolishing illegal structures. The comprehensive land use plan for Malay municipality must legally secure approval from the Aklan Provincial Land Use Committee, which includes HLURB as member, yet approval has not been secured since 2011.

The interpretation of beachfront setbacks as a first step in planning interventions with regards to the real, ongoing threat of sea level rise as well as gridlock on the approval of the comprehensive land use plan (CLUP) indicates a challenging political environment for all stakeholders, especially for the local government of Malay. One of the many requirements for compliance for all municipal planning departments is the mainstreaming og both the Disaster Risk Reduction and Management Act (RA10121) and the Climate Change Adaptation Act. The same national-level official said:

There was previously a zoning ordinance for Boracay, but I could not understand why it is not being followed. Even the agricultural land is now being bulldozed for all these buildings. It is really critical that the PLUC reviews all these plans and actual implementation, and that includes if climate change adaptation and disaster risk reduction is mainstreamed into the CLUP. Despite the RA10121 and the Climate Change Act being relatively new laws, all municipalities are required to supplement and enhance existing CLUPs with additional chapters for climate and disaster vulnerability assessments (Govt_RL, 2014).

Stakeholder groups are indicated in capital letters while the resulting interview themes are listed in normal letters.

**Table 2.**Degree of similarity between responses recorded between and among themes and source classifications/stakeholders related to their Boracay environment (excerpt).

| Node/Source A | Node/Source B | Pearson Correlation Coefficient |
|---|---|---|
| Beach erosion | Awareness of local environments | 0.68874 |
| Beachfronts and foreshores | Awareness of local environments | 0.81222 |
| Beachfronts and foreshores | Beach erosion | 0.75924 |
| Building and construction | Beachfronts and foreshores | 0.67659 |
| Congestion | Awareness of local environments | 0.56492 |
| Congestion | Beachfronts and foreshores | 0.54615 |
| Disasters, risk reduction, and management | Awareness of local environments | 0.48153 |
| Disasters, risk reduction, and management | Beach erosion | 0.18238 |
| Sea level rise | Beach erosion | 0.61754 |
| Sea level rise | Perception of tourism change in the island | 0.53465 |
| Sea level rise | Beachfronts and foreshores | 0.67515 |
| Environmental degradation | Awareness of local environments | 0.73105 |
| Environmental degradation | Beachfronts and foreshores | 0.72624 |
| Environmental degradation | Building and construction | 0.63284 |
| Environmental degradation | Beach erosion | 0.57581 |
| Environmental degradation | Congestion | 0.57146 |
| NON-GOVERNMENTAL ORGANISATIONS | Environmental management | 0.70097 |
| RESIDENTS | Environmental management | 0.85146 |
| TOURISM BUSINESS | Environmental management | 0.80588 |
| GOVERNMENT | Environmental management | 0.89689 |
| VISITORS | Environmental management | 0.91033 |
| VISITORS | Illegally built structures | 0.67238 |
| TOURISM BUSINESS | Perception of tourism change in the island | 0.85755 |
| RESIDENTS | Perception of tourism change in the island | 0.87150 |
| Perception of tourism change on the island | NON-GOVERNMENTAL ORGANISATIONS | 0.68496 |
| Perception of tourism change on the island | Infrastructure development | 0.79198 |
| Perception of tourism change on the island | Beachfronts and foreshores | 0.78171 |
| Perception of tourism change on the island | Congestion | 0.70275 |
| Perception of tourism change on the island | Illegally built structures | 0.69995 |

The planning interventions require rapid implementation, and some interview respondents regard national laws on disaster risk reduction and climate change adaptation as necessary due to Boracay's rapidly degrading environment (see Table 3). Stakeholders are almost unanimous in saying that Boracay's environment requires drastic management measures to maintain its status as a sustainable tourism destination, with visitors as the stakeholder group registering the highest similarities in their responses regarding environmental management (Pearson coefficient 0.910329).

Tourism development is the main driver of environmental change on Boracay, which has been subject to a process of rapid urbanisation. This theme emerges strongly in the analysis of the stakeholder interviews and is evident in the comments stakeholders make in connection to infrastructure development, the island's beaches and foreshores, traffic congestion, and the proliferation of illegal construction (see Table 2). The only plausible reason for this perception of change is that the island's economy is entirely dependent on tourism. This is in keeping with the TALC model's stagnation stage, in which significant deterioration of tourism assets and the breeching of carrying capacities inhibits further growth and necessitates stakeholder intervention, especially by government (Butler, 1980). Recent studies have highlighted that the island may already have entered TALC's decline stage, meaning that, absent intervention, it will not become the kind of tourism metropolis predicted by some models (Smith *et al.*, 2011). This active involvement from the public sector and various stakeholders necessitates intensive political negotiation as tourism development will only be able to make a positive contribution to the island without further social, economic, and environmental degradation when current stakeholder relationships are evaluated and acted upon (Butler, 2006). The island's political, social, economic, and natural conditions will be further tested and indeed are already being challenged by various climate-related disasters occurring around the country, the most recent of which was Typhoon Haiyan in 2013.

**Figure 3.** Tourism has transformed two-thirds of the island into dense urban spaces. This aerial photo shows the northern rural barangay (district) Yapak and the central urban *barangay* Balabag (Author's Fieldwork, July 2013).

Some of the first migrants to establish their own resort and tourism businesses as well as repeat visitors have observed the significant changes occurring on Boracay. A husband-and-wife team, who settled and are now operating a small accommodation business, recalls:

We've been living here on the island most of the time since 1986. There's been a lot of changes since then. There was no electricity on the island, no cement roads, and there were just only three or four delivery tricycles. The road was narrow, muddy, and rough, same location. It was just about two meters wide. I arrived in February of 1986 [...] after we were stuck in Manila because of the People Power Revolution (Resident&TourismBiz_JDT&JCT, 2013).

Stakeholders who have been intimately familiar with the island since the 1980s and 1990s have also witnessed the social changes brought by tourism, which have been experienced by similar island resorts (Picornell, 2014). The increasing number of tourism establishments and subsequent urbanisation has contributed to congestion and crowding:

In the early '90s when I first came here, it wasn't that congested with establishments. There were no multinational firms engaged in the tourism businesses, mostly local businesses. There are now more clubs, and I've seen prostitution openly. A lot has changed. It's just like Ermita and Makati with sand. I'm not a fan of the island already, we're just here because we're organising the tournament for my cousin, who's based abroad, and we're not spending our trip here (Visitor_PJS&ADR, 2013).

I started living here in 1996 when I was 16 years old. I graduated from high school on the island. I remember that Boracay wasn't crowded then. There was no electricity. We used gas lamps for lighting at night. My mother was working at Marimar Resort then, and all the resorts were made of local, native materials. It was really rustic. A majority of the tourists then were Japanese, unlike now where it's dominated by Koreans. Buildings have sprung up all over now (Resident&TourismBiz_MW, 2013).

Adding to the crowding and congestion is the influx of labour for various tourism-related establishments as well as the families of migrants. The local government has legislated to place controls on the hiring of workers and labour migration, especially for construction workers:

One of the major problems is migration to the island from other islands, especially construction workers, who will also bring their family members. After the construction project is over, they remain and will usually just live in informal dwellings while they find more work. Jobs intended for locals are currently prioritised under Municipal Ordinance 6040, and population control is linked to jobs management within the island (GOVT&Resident_FDLS, 2013).

The concentration of both locals and migrant workers on the island, together with tourist visitors, compared to the rest of Malay municipality has made Boracay one of the most densely populated areas in the municipality. Despite Boracay's reputation as an island idyll, two of the island's three *barangays* (villages or districts) are now considered urban. Although Malay town on the mainland is the municipal capital, Boracay today accounts for three-fifths of Malay's population and wields tremendous political and economic power over the rest of the municipality. Yet densely populated small islands are also particularly prone to conflicts over land use (Salvacion & Magcale-Macandog, 2015). Such is the importance of the island that the municipal government has field offices on Boracay, with major officials regularly commuting from the town hall on the mainland. This tourism-driven upsetting of traditional island-mainland centre-periphery relationships (Grydehøj, 2014; Grydehøj *et al.*, 2015) can render governance more complex.

**Table 3.** Malay municipality's rural-urban population (2010).

| Location | Urban Districts | Rural Districts | Population | % of Total |
|---|---|---|---|---|
| Boracay Island | 2 | 1 | 28,369 | 61.9% |
| Mainland Malay | 2 | 12 | 17,442 | 38.1% |
| Totals | 4 | 13 | 45,811 | 100% |

The immense political and economic power of the island's three districts is not lost on stakeholders. Tourism businesses in particular are critical of the perceived cross-subsidy by which the island's tax revenues are shared among the rest of the municipality's other 14 districts. A long-time migrant and established accommodation property owner as well as a long-time resort manager echo the same thoughts:

I think it's unfair that the revenues of Boracay are used in other areas of Malay.Boracay should become a municipality and split from Malay. I'm not in favour of making Malay a city. Malay is just a small, sleepy town. Where will they build a city? There's no established town proper where you can build into a cityproper, and commercial establishments (Resident&TourismBiz_JDT&JCT, 2013).

They want to subdivide Aklan into two congressional districts, which is convenientfor political families. But converting Malay into a city is problematic because of the land area requirement, so I think it wouldn't qualify. But if you ask me,Boracay is Boracay, and Malay is nothing but a sleepy town. Most of the revenuesof Malay, about 80-90%, come from the island (TourismBiz_JC, 2013).

The island is also a base for launching political campaigns, especially in local elections:

If you win in at least two barangays/districts here during elections, you'll win in the elections in the entire Malay municipality. The island is well populated and rich in votes. I still strongly suggest that Boracay becomes a municipalityrather than that Malay becomes a city (Resident&TourismBiz_JDT&JCT, 2013).

In contrast, however, some contributors see the potential cityhood of the town of Malay, from which the island is politically and administratively governed, as a pathway to better services, more independence from the provincial government, and a more capable local government, responsive to the needs of the town's stakeholders. Achieving city status for Malay municipality would allow it to rival the provincial capital of Kalibo, which is similarly seeking city status. A high-ranking local government official, who is also a Boracay resident, states:

We foresee Malay becoming a city of which Boracay is a part. We already fulfil the criteria for population size, annual income, and land area. Since the municipality has an island, it is exempt from the size requirement. We're preparing it in our Annual Investment Plan to have Malay cityhood legislation be passed in the Philippine Congress. However, we should start preparations within our local government unit (GOVT&Resident_FDLS, 2013).

Though the cityhood option could be a welcome political exercise to pursue, a long congressional campaign should not replace good governance and control of tourism development and investment on the island to manage the island's adaptive capacities to urbanisation and by extension to climate change adaptation.

Stakeholders who are privy to new rules in tourism investment regulation on the island concede that these have already been imposed to control construction activities since the island's infrastructure is being stretched to its limits. A 20 million peso (AU$604,000) minimum investment rule for new tourism establishments has been passed by the municipal council. Despite this, there are reports that monitoring of these construction permits is so weak that some establishments proceed with expansion or construction regardless: Compared with the money earned from the investments, the fines imposed by the municipality are puny, described by one stakeholder as "loose change" (Govt_RC, 2013).

Members of the tourism industry are supportive of the move to impose a minimum investment rule as legislated by the municipal council since this would discourage the establishment of run-of-the-mill resorts and would prevent competition with existing local businesses. The industry group Boracay Foundation Inc. (BFI) says:

The purpose of the moratorium is really to stop construction work that did not go through the proper processes, like shanties that have sprung up like mushrooms. The point is that, if you are going to invest and you are an outsider, you had better invest big since if you are just going to invest a small amount, you will be competing with the locals, and the locals can invest that amount themselves (TourismBiz&NGO_DS&PM, 2013).

This industry group also supports the local government moratorium on building andconstruction within 15 meters of the island's main road (Municipal Executive Order 25), effective 1 October 2014 (Interaksyon, 2014). This moratorium pales in contrast with Proclamation 1064 issued by Office of the President Gloria Arroyo in 2006, which imposes a "15-meter buffer zone on each side of ALL existing roads and trials that are reserved for right-of-way and shall form part of the area reserved for forestland protection purposes" (Macapagal-Arroyo, 2006).

These conflicting legal statutes set forth by the Philippine national government and the Malay local government are a manifestation of the legal quandaries faced by tourism stakeholders, which must fight in the courts when confronted by legal challenges. It is also a manifestation of the lack of foresight and poor integration of planning legislation across different levels of governance. This legal conflict simply transfers the burden to the courts, which could involve years of litigation, when itwould have been more practical to 'do it right the first time'. Climate change adaptationwould require fine-tuned approaches to the legislation and implementation of existing and future laws impacting on the capability of planners and stakeholders to undertake required action. Such stalemates do not help efforts in this policy direction.

# 4. STAKEHOLDER VIEWS AND REALISATIONS REGARDING CLIMATE CHANGE AND CLIMATE CHANGE ADAPTATION

One objective of the present research is to qualify the various stakeholders'awareness of issues concerning climate and climate change, tourism, and governance.The coded responses made by the various stakeholder groups indicate that climate change awareness and climate change adaptation tend to be strongly associated with one another (see Table 4). A couple who have been living on the island and who have owned and operated a small hostel since the 1980s have noted:

Whenever there were typhoons, we experienced before they were very strong, unlike now. So if you're talking about climate change, there have really been changes in the typhoon patterns. I really believe that there's climatic change because usually the onset of the rainy season is the third week of May. There were very distinct times when there is a change from the *amihan* (northeast monsoon) to the *habagat* (southwest monsoon). Now you can't tell if there will be a change in the monsoons, and the rainy season is getting shorter. The rains have also become less predictable than when I first came here. There was never any rain during the Holy Week or Easter period. Now, it rains even during Holy Week when it should be hot and dry (Resident&TourismBiz_JDT&JCT, 2013).

The local government sector, at both the municipal (Malay) and provincial (Aklan) levels, is cognisant of the management challenges accompanying the realisation of climate change as a policy issue. The passage of the Climate Change Act has prompted various levels of government to initiate steps toward adaptation and to integrate these into existing planning processes:

The province was able to produce a climate change map, especially those that will affect land use, including a hazards map from the Mines and Geosciences Bureau. The Provincial Land Use Committee is comprised of national agencies, which include the departments of Agrarian Reform, Public Works and Highway, Environment and Natural Resources, Tourism, among others. The Provincial Planning and Development office is the secretariat. The provincial land use plan is due for an update as its validity is from 2008-2013, and we are in the process of collecting data. The mandate is now therefore to update it with climate change adaptation in one of the chapters of the provincial land use plan (Govt_RE, 2013).

Whether or not these resulting plans serve their purpose of guiding and providing appropriate action for adaptation remains to be seen since, at the time of writing, there has never been a land use plan that has been compliant with the new guidelines of the Housing and Land Use Regulatory Board (HLURB), which was implemented in 2014. Local governments are required under the Republic Act 10121 (Philippine Disaster Risk Reduction and Management Act of 2010) and the Climate Change Act to integrate climate change adaptation and disaster risk reduction into their comprehensive land use plans. The current practice is for climate change adaptation offices to be combined with municipal disaster risk reduction and management offices, which are tasked with undertaking CCA/DRR plans. However, there is currently only one pilot municipality in which CCA, and disaster risk reduction is still in the process of being integrated into the comprehensive land use plan (Govt_RL, 2014).

The introduction of climate change adaptation and disaster risk reduction into a governance system in which planning technology and capacity remain in their infancy would be a budgetary if not an internal staffing capability problem for many local government units. Major cities and higher-income municipalities would have no difficulty securing funding for what could be an expensive planning exercise, especially if the local governments procure the services of consultancy firms. Malay municipality, Boracay's parent government unit, has hired external consultants for both the municipality's comprehensive land use plan and the island's tourism master plan. Despite this, the municipality has failed to actually submit a functional and integrated CLUP to the Aklan Provincial Land Use Committee. One of the national oversight agencies has noted that:

Malay and Boracay hired consultants for the CLUPs for their respective areas, so we in particular don't provide any technical assistance to the municipality. The officials of Malay attended all four of the training modules for CLUP, and they have been able to produce outputs for each of the modules, but they have not submitted a consolidated CLUP (Govt_RL, 2014).

A hotel manager who was invited to one of the consultative planning sessions for the municipality's comprehensive land use plan noted that the municipality presented a vision for its tourism future:

In 2011, the arrivals were about 900,000, in 2012 1.2 million, and in 2013 they are targeting 1.3 million. I attended the comprehensive land use plan public presentation, and they are really expecting Boracay to accommodate more guests. They have an idea for the future, but the details seem lacking. We really don't have the infrastructure at the moment that could keep pace with the arrivals of travellers (TourismBiz_RA, 2013).

That the municipal government identifies tourism as a foundation for its comprehensive land use planning says much about the importance of this economic sector. Both the municipal and provincial governments feel that they need support for climate change adaptation:

The provincial government of Aklan still considers agriculture and tourism as the twin engines of growth for the province, and we hope the national government will continue to show support to help us protect the island of Boracay. Preservation of the environment and climate change adaptation will be a priority, and we hope that we will be able to collaborate with other stakeholders to help us prepare (Govt_RE, 2013).

The thousands of severely understaffed and technically deficient municipal planning and development offices (MPDOs) in particular require support following the recent publication of the 'Supplemental Guidelines on Mainstreaming Climate and Disaster Risks' in the Comprehensive Land Use Plan in 2014 (PCCC, 2014). At the time of fieldwork, one of the resource workers from the Housing and land Use Regulatory Board (HLURB) emphasised that the process of preparing CLUPs had become more stringent:

We're preparing a supplemental guidebook for climate change adaptation mainstreaming aside from the existing CLUP guidebooks, and it will be completed soon. The process of getting CLUPs approved has gotten more stringent since they're now required to have chapters on climate change adaptation in their CLUP submission to the provincial councils for approval. According to the Climate Change Commission, we're the third-most vulnerable country, and we see that it's through the CLUP and the Zoning Ordinance that mainstreaming will be undertaken. It's through the CLUP that disaster-prone areas will be identified, and the implementation tool will be the zoning ordinance(Govt_RL, 2014).

This respondent also recognised the limited capability of local governments within the Western Visayas region (where Malay, Aklan is located) even to update their lapsed CLUPs. The burden of integrating all legislation affecting land and water use planning in individual municipalities (and in the case of the parent provinces, of integrating the municipalities' various CLUPs into a cohesive Provincial Physical Framework Plan), combined with the multiple work functions of planning officers (such as taking on the role of agriculture or environment officer) is simply too daunting for municipal and provincial governments. This is a common quandary repeated regions across the country. This same respondent gives a sobering account of the state of completion of CLUPs in her region:

These multiple functions of municipal planning officers hamper the completion of the mandated functions of their offices. In Western Visayas, only 10% of municipalities were able to have their municipal comprehensive land use plans approved by their respective provincial land use committees and provincial councils. This is one of the reasons why they couldn't complete their CLUPs as well as numerous laws and projects that need to be implemented, like bottom-up planning [...]. The mayor presses their planning offices to produce a CLUP since investors will be looking at these documents as well. Under RA10121 and the Climate Change Act, they are required to integrate climate change adaptation into their land use plans. They don't have additional budgets for that purpose. The Climate Change Commission is just piloting the project of integrating into CLUP in just one town in Davao (Govt_RL, 2014).

The regional HLURB office is nevertheless critical of Malay municipality's failure to update its CLUP despite the support extended to it from all levels of government (Govt_RL, 2014).

Since Boracay has a special status as the country's premier small island tourism destination, it is also required to produce an updated Tourism Master Plan, with support from the national government through the Department of Tourism and its affiliated agency, the Tourism Investment and Enterprise Zone Authority (TIEZA), which is tasked with regulating tourism infrastructure projects. Even TIEZA, however, had no one heading the field office in Boracay at the time of fieldwork (TourismBiz&NGO_DS&PM, 2013).

There is a general observation among planning professionals in the Philippines that local governments face capacity challenges in terms of formulating the basic comprehensive land use plans (CLUP) and the more specific tourism, disaster risk reduction, and climate change adaptation plans. This is despite the fact the Philippines is considered to be one of the earliest adapters of local government decentralisation in Southeast Asia, through the legislation of the Local Government Code of 1991, and is now on its third decade of implementation (Gomez & Buenaventura, 2012).

## 5. STAKEHOLDER GROUP RELATIONSHIPS AND INTERACTIONS: FEELINGS OF DISTRUST AND HELPLESSNESS?

The Philippines is currently considered as having instituted multi-stakeholder participation in its planning processes, ancestral domains, conservation of protected areas, and legislation (Bryant, 2000). This is well documented among members of the various stakeholder groups. Although stakeholder participation events are rife with dissent and boisterous objections, as can be expected in a free-wheeling democracy like that of the Philippines, it is important to note that these stakeholder consultations do take place on Boracay. This can be considered an important step toward enhancing the stakeholder capacities, not only in building confidence but also in building trust between these groups. Trust among stakeholders is vital for "the effective implementation of climate change adaptation and mitigation measures" on small islands (Petzold, 2015).

Local government autonomy and delineation of authority, especially between the Aklan provincial government and the Malay municipal government, under which Boracay is governed, can be described as a fine line between 'respectful distance among various governance levels' and 'an urgent push to implement reforms'. An environment officer in the provincial government mentions that although there exists a national-to-provincial multilevel agency for discussing environment and tourism issues on the island, the ultimate decision on implementing any policy recommendation remains with the Malay municipal authorities:

**Table 4**. Degree of similarity between responses recorded between and among themes and source classifications/stakeholders relating to climate and climate change and related pairings (excerpt).

| Node/Source A | Node/Source B | Pearson Correlation Coefficient |
|---|---|---|
| **Climate change awareness** | Climate change adaptation | 0.915656 |
| **GOVERNMENT** | Climate change adaptation | 0.69343 |
| **GOVERNMENT** | Climate change awareness | 0.666593 |
| **Inter-agency coordination** | Climate change adaptation | 0.644456 |
| **Legal issues** | Climate change adaptation | 0.60695 |
| **Local awareness of place in climate terms** | Climate change awareness | 0.513703 |
| **Local awareness of place in climate terms** | Indigenous beliefs on climate | 0.509909 |
| **Local government staff qualifications** | Climate change adaptation | 0.402636 |
| **Local livelihoods** | Climate change awareness | 0.283162 |
| **Local livelihoods** | Climate change adaptation | 0.264363 |
| **National-municipal government relationship** | Climate change adaptation | 0.669718 |
| **National-municipal government relationship** | Disasters, risk reduction, and management | 0.633226 |
| **National-municipal government relationship** | Climate change awareness | 0.589536 |
| **NON-GOVERNMENTAL ORGANISATIONS** | Climate change awareness | 0.41717 |
| **NON-GOVERNMENTAL ORGANISATIONS** | Climate change adaptation | 0.392468 |
| **Perception of tourism change on the island** | Local awareness of place in climate terms | 0.504304 |
| **Perception of tourism change on the island** | Climate change adaptation | 0.441568 |
| **Political interference** | Climate change adaptation | 0.314891 |
| **Power relations** | Climate change adaptation | 0.344137 |
| **Power relations** | Climate change awareness | 0.31862 |
| **Provincial-municipal government relationship** | Climate change adaptation | 0.581588 |

**Table 5.** Degree of similarity between responses recorded between and among themes and stakeholders related to stakeholder interactions, relationships, and related themes (excerpt).

| Node/Source A | Node/Source B | Pearson Correlation Coefficient |
|---|---|---|
| National-municipal government relationship | GOVERNMENT | 0.864895 |
| NON-GOVERNMENTAL ORGANISATIONS | GOVERNMENT | 0.654488 |
| NON-GOVERNMENTAL ORGANISATIONS | Inter-agency coordination | 0.595451 |
| NON-GOVERNMENTAL ORGANISATIONS | Inter-agency coordination | 0.595451 |
| Perceptions of governance | Local governance | 0.888281 |
| Perceptions of injustice and inequality | NON-GOVERNMENTAL ORGANISATIONS | 0.546186 |
| Perceptions of injustice and inequality | GOVERNMENT | 0.535264 |
| RESIDENTS | GOVERNMENT | 0.818831 |
| RESIDENTS | NON-GOVERNMENTAL ORGANISATIONS | 0.726217 |
| RESIDENTS | National-provincial government relationship | 0.472716 |
| TOURISM BUSINESS | RESIDENTS | 0.95221 |
| TOURISM BUSINESS | NON-GOVERNMENTAL ORGANISATIONS | 0.674676 |
| TOURISM BUSINESS | National-municipal government relationship | 0.613966 |
| TOURISM BUSINESS | Provincial-municipal government relationship | 0.448283 |
| TOURISM BUSINESS | National-provincial government relationship | 0.395063 |
| Tourism business operators' industry linkages | RESIDENTS | 0.651687 |
| Tourism business operators' industry linkages | TOURISM BUSINESS | 0.637258 |
| Tourism business operators' industry linkages | GOVERNMENT | 0.444164 |
| Tourism business operators' industry linkages | NON-GOVERNMENTAL ORGANISATIONS | 0.421259 |

We've also set up Bantay Boracay, which is a multi-government agency to discuss sectoral issues that are related to Boracay. The outputs of these discussions are made with the view of forming policies, and possibly resolutions. The one-entry, one-exit policy was a result of this coordination, where boats will no longer dock in front of the main White

Beach but instead at the jetty ports. This is only a resolution but would have to be implemented by the Malay local government as the provincial government doesn't want to infringe on the autonomy of the local government of Malay (Govt_JKA, 2013).

Other government agencies at the national level are even more critical, saying that local governments have made no progress since 1991 in assuming the roles that the national government have delegated to municipal and provincial governments in tourism business accreditation and standards, land titling, environmental management, and urban planning. The regional tourism office has commented on municipal tourism management:

In 1991, when the Local Government Code was enforced, the local government naturally took over. We used to issue licenses and regulations until that point. Now we only do accreditation, and the business licenses are given by the local government. Until that point, everything went haywire since everything is now devolved to the local government. The island was previously managed by the Philippine Tourism Authority, now reorganised as the Tourism Investment and Economic Zones Authority since it is the infrastructure arm of the DOT (Govt_RC, 2013).

Another major resort manager has given insight into current land titling, land-grabbing, and indigenous resident affairs:

Most of those who have developed resorts, approximately 80-90% here, only have tax declarations, not land titles. The land titling process here is going to be very messy and contentious due to overlapping claims, and most of them will fight for their land rights since they have already developed the land with their resorts and properties (TourismBiz_JC, 2013).

When then asked how this particular observation on local governance could affect efforts to mainstream climate change adaptation in tourism on the island, this resort manager adds:

Climate change adaptation should really be implemented by the local government with sectoral representation from the national government, depending on the sector - DOT for tourism, DENR for environment, DA for agriculture, *etc.* We just couldn't implement anything without any coordination with the local governments, like requests for training, capacity building, *etc.*(Govt_RC, 2013).

The gridlocked dynamics and the keen observance of turf between the variouslevels of local governance are not lost on members of the tourism business communityand could partly explain why governance has become a burden for their operations. With very specific mandates to implement their official charters, a hotel operations manager observes that "each government office has its own ideas on how to manage and improve Boracay, but there seems to be no structure in how organisations coordinate their efforts" (TourismBiz_RA, 2013; *cf.* Andexlinger, 2015).

Land acquisition, registration, and the relationships between the government, tourism businesses, and the indigenous residents are among the most contentious issues and have even occasionally resulted in violence. Land disputes on the island regularly make it to national headline news. There is the existing conflict over Friday's Resort between relatives of the incumbent mayor and the Manila-based property company Boulevard Holdings, where half of the resort was closed and the other half of the property remained in operation but was effectively paralysed because of the dispute, despite the case having been resolved in the courts in 2009 (Dumlao, 2014).

The intertwined influence of tourism businesses, the regulatory weakness of the municipal government, and slow legal processes for resolving disputes lead tofurther erosion of trust and confidence by other stakeholders in the local government'scapability to undertake the necessary measures to confront Boracay's numerous environmental, political, and social challenges. The provincial government, for its part, emphasises that the root cause is the island's rapid and unmanageable urbanisation, as manifested in perceptions of overcrowding:

The problem really is that private sector development has overtaken the local government regulation, and oversight is now very difficult. Government has fallen so far behind the regulation process that, even if the national and local government agencies acted, it is now difficult to change current development. The private business sector has the resources and money to go to court and secure temporary restraining orders [TROs] and injunctions to block government initiatives. We also resort to going to court to settle issues with some stakeholders, like road right-of-way, as this is sometimes the only way to settle disputes. The provincial government only serves as support as the municipality of Malay is on the frontline when it comes to regulatory oversight functions for the island of Boracay (Govt_RE, 2013).

In the absence of an updated comprehensive land use plan (CLUP) for the municipality and a useable tourism master plan, the private tourism business sector has become the *de facto* 'urban designer' for the island, with the municipal processes simply becoming rubber stamps for the required planning permissions, which will be granted regardless since planning and development controls are very feeble. This has, it is felt, led to haphazard and unintegrated development (*cf.*Pigou-Dennis & Grydehøj, 2014). This is the opinion of the provincial government, which approves all municipal CLUPS and the plethora of other planning documents required by national laws (disaster risk reduction and management, tourism, strategic agro-fisheries development zones, *etc.*). This opinion is very much aligned with the findings of Gavin Shatkin (2008), who notes that the urbanisation that has taken place in Metro Manila is unprecedented in terms of the extent of "privatization in urban and regional planning," with large property companies leading the charge for urban development. This also fits perfectly into the general lack of local government capacity for undertaking planning activities with the support of civil society, thereby shielding decision making from powerful economic interests (Shatkin, 2000). This forces the need to revisit the linkages and power relationships between and among various tourism stakeholders (Maguigad, 2013).

Representatives of a non-governmental organisation advocating for the island's indigenous Ati population are scathing about the local government. Their views during the researcher's July 2013 interviews border on verbal hostility and exasperation. This is understandable since the main spokesperson for the Boracay Ati Tribal Organisation (BATO), Dexter Condez, was fatally shot by a security guard for one of the island's big resorts in February 2013. Many believe the shooting to be related to the 2.1 hectare prime piece of beachfront property that was awarded to BATO by the national government in a writ of possession through a Certificate of Ancestral Domain Title in April 2012 (2013). Dexter Condez was among the few Ati who were able to integrate into mainstream Aklanon society and was engaged in the indigenous arts. He was thus able to communicate with ease with various other stakeholders (Angan, 2013). His killing was portrayed in the media as the "ugly face of Boracay development that is rooted in artifice, lack of cultural rootedness that has oppressed the Ati indigenous people because of their black skin and lack of formal education" (Alegre, 2013).

The interviewed BATO representative was aware of the existing 25+5 meter beach setback rule but was quick to mention that Mayor John P. Yap was in violation of the local government-imposed ordinance:

**Figure 4.** Boracay struggles with the pace of urbanisation fuelled by uncontrolled tourism development and complicated by governance gridlocks at the local, provincial, and national levels (Fieldwork Photographs, August 2013)

Our land here has also been affected by the 25+5 meter rule of beachfront setback. It's the reason why this is how our village looks since we are sandwiched by the road and the beachfront. However, the mayor has a business establishment named Summer Place that also violates the imposed regulations. I will applaud the mayor if he will initiate the demolition against his own establishment as well as all those businesses in violation of the rules (NGO_LT, 2013).

The BATO representative even highlighted the glaring disparity between the relationship BATO has with the national government through the National Commission for Indigenous Peoples (NCIP) and BATO's relationship with the local government:

We don't have a good relationship with the Malay local government, but we have an excellent relationship with the national government through the National Commission for Indigenous Peoples. We're not involved and were disregarded by the local government in the local planning of the island. It's only the national government that continues to care for us through the NCIP (NGO_LT, 2013).

When prompted to answer questions regarding climate change adaptation, her tone was mixed:

I have to be silent on legal matters. You have to ask the government about climate change since they're the ones who will be managing that. If they won't do anything about it, we'll have to depend on ourselves. Our world is interconnected, so we have to cooperate with whatever initiatives they will be doing. But even if they're that bad to us, we still love them since what they do will eventually get back to them. We have about 32 families with about 200 people living in the village. Dexter has been the only one killed so far in our village.If they need a representative for consultations, they usually contact Delsa Justo, our chieftain. If there are seminars related to education, I also usually attend (NGO_LT, 2013).

The self-reliant tone ("we will have to rely on ourselves") reflects the fatalist dimension of Cultural Theory's 'Five Ways of Being', which is typically expected from disempowered and marginalised sections of society, which take comfort in the "rationality of their powerlessness" (Forsyth, 2003). It is interesting to note, however, that this respondent uses the words "we still love them" and the concept of karma (or the Golden Rule), further indicating that her organisation will participate in public consultations with the government when its presence is requested. According to the Ati interviewee, this conciliatory tone, despite the scathing criticism, is a reflection of the religious teaching that the Ati community receives from the Catholic Church through the Daughters of Charity of St. Vincent de Paul, which provides education, training, and food (NGO_LT, 2013).

These governance gridlocks as well as the strained stakeholder relations do not augur well for a concerted effort to mainstream climate change adaptation. Cross-cutting interaction from all stakeholders is needed to create critical momentum to go a step further from the broad adaptation plans towards a plan attuned to an island focused on tourism.

## 6. CONCLUSION

Tourism development has transformed and driven Boracay from a sleepy tropical island to the Philippines' top small island tourist destination over a span of more than three decades, supporting the development of Malay municipality and the remainder of Aklan province. The urbanisation that resulted from this development has created a political ecology that is extremely problematic in the terms of strained relationships between the various tourism stakeholders, pitting the strong tourism business groups against a politically weak local government hobbled by conflicts of interest in fulfilling national legal and comprehensive land use planning guidelines. The local government is weighed and found wanting in communicating to the various stakeholders the need for unity on single-issue debates. It would be difficult for an adaptation plan for tourism and climate change to be pursued as the stakeholders are united primarily in distrust and helplessness. More local independence in fiscal and governance terms by converting the municipality into Aklan's province's first city may provide some hope for improving planning and adaptation capacities. However, the success of such a conversion to cityhood would still ultimately be dependent on the relationships built over the years among the stakeholders, which could prove much more difficult to mend than preparing an actual island tourism climate change adaptation plan. The cornerstone for achieving adaptation success is therefore for the local government to lead an effort to bridge gaps between stakeholders (especially between the tourism businesses, indigenous peoples, and resident population) to ensure proper ownership of any planning and climate change adaptation plans.

The planning process to mainstream climate change adaptation into the island's tourism sector is challenged by various factors aside from the existing quality of the stakeholders' relationship, their political ecology. First, the comprehensive land use plan (CLUP) for Malay has not passed the muster of evaluation by the Aklan Provincial Land Use Committee. CLUPs are the core plans upon which all sectoral plans (including tourism, climate change adaptation, and disaster risk reduction plans) are ultimately based. This is the most vital process that must be accomplished and requires an honest engagement with the stakeholders, not just the kind of tokenistic public consultation that many stakeholders report having experienced (*cf.* Pugh, 2013; Grove & Pugh, 2015). Climate change and climate-related disasters will impact the livelihoods of many who are dependent on the tourism industry. Beach erosion is already occurring on parts of the island and is expected to become more pronounced with the effects of sea-level rises, compounded by seasonal typhoon and monsoon seasons. These and other manifestations of climate change will bring challenges and opportunities that can only be adequately addressed when the existing political ecology is integrated into the planning process.

This research has demonstrated the novel use of political ecology in tourism and climate change adaptation as a means of improving planning processes in the Philippines. The use of this approach considers a very important aspect of environmental planning – that planning processes are not only inherently the domain of technical experts working in an apolitical environment but also that they are inherently political. This is not hard to implement as existing local government codes require intensive, if not honest, stakeholder participation. This approach will ensure proper responses, which will benefit not only islands with tourism industries but also other local government units with similar tourism potentials.

## REFERENCES

1. ABS-CBN News ( 2013) 'Boracay's White Beach Voted among the World's Best',*ABS-CBN News*. Available at http://abs-cbnnews.com/lifestyle/02/28/13/boracays-white-beach-voted-among-worlds-best. Accessed 20 October 2014.

2. Alegre, J.D. (2013) 'Killing of Indigenous Leader Bares Ugly Face of Boracay Development', *Philippine Daily Inquirer*. Available at

3. http://lifestyle.inquirer.net/97881/killing-of-indigenous-leader-bares-ugly-face-of-boracay-development. Accessed 20 October 2014.

4. Andexlinger, W. (2015) 'Suburban Processes Of Islandisation In Austria: The Cases Of Vienna And Tyrol', Urban Island Studies, 1, 118-133.

5. Angan, J. (2013) 'Beyond the Beach: The Untold Story of Boracay's Ati Tribe',*GMA Network News*. Available at

6. http://gmanetwork.com/news/story/313920/lifestyle/artandculture/beyond-the-beach-the-untold-story-of-boracay-s-ati-tribe. Accessed 20 October 2014.

7. Arnaldo, M.S.F. (2012) 'DOT Launches P265-Billion National Tourism Strategy',*Interaksyon*. Available at http://interaksyon.

com/article/30982/dot-launches-p265-billion-national-tourism-strategy. Accessed 8 December 2012.

8. Broderick, J. (2009) 'Voluntary Carbon Offsetting for Air Travel' in S. Gössling & P. Upham (eds) *Climate and Aviation.* London: Earthscan.

9. Brown, D., C. Taylor, G. Baldy, G. Edwards, & E. Oppenheimer (1990) 'Computers and QDA: Can They Help It? A Report on a Qualitative Data Analysis Programme', *Sociological Review*, 38(1),134-150.

10. Bryant, R.L. (2000) 'Politicized Moral Geographies: Debating Biodiversity Conservation and Ancestral Domain in the Philippines', *Political Geography*, 19(6), 673-705.

11. Butler, R.W. (1980) 'The Concept of a Tourist Area Cycle of Evolution: Implications for Management of Resources', *Canadian Geographer*, 24(1), 5-12.

12. Butler, R.W. (2006) 'The TALC and Local Involvement' in R.W. Butler (ed.) *The Tourism Area Life Cycle Vol. 1: Applications and Modifications.* Clevedon: Channel View.

13. Carter, R.W. (2004) 'Implications of Sporadic Tourism Growth: Extrapolation from the Case of Boracay Island, the Philippines', *Asia Pacific Journal of Tourism Research,* 9(4), 383-405.

14. Dasgupta, S., H. Brecht, B. Laplante, S. Murray, & D. Wheeler (2012) 'Sea-Level Rise and Storm Surges: High Stakes for a Small Number of Developing Countries', *Journal Of Environment & Development,* 21(1), 120-138.

15. DOT (2012) *Department of Tourism Regional Distribution of Travellers in the Philippines 2011-12.* Manila: Department of Tourism.

16. DOT & WTO (1991) *Philippine National Tourism Master Plan 1991-2010.* Manila: Philippine Department of Tourism & World Tourism Organization.

17. Dubois, G., P. Peeters, J.P. Ceron, & S. Gössling (2011) 'The Future Tourism Mobility of the World Population: Emission Growth Versus Climate Policy', *Transportation Research Part A,* 45(10), 1031-1042.

18. Dumlao, D.C. (2014) 'Fridays Boracay in Crisis Due to Land Dispute', *Philippine Daily Inquirer.* Available at http:// business. inquirer.net/167573/fridays-boracay-in-crisis-due-to-land-dispute. Accessed 23 November 2015.

19. Forsyth, T. (2003) *Critical Political Ecology: The Politics of Environmental Science.* London: Routledge.

20. Gomez, J.E.A., & K.V. Buenaventura (2012) 'Fields of Gold: Empowering Local Officials to Catalyse Rural-to-Urban Development in Tarlac Province, the Philippines', *Space and Polity,* 16(1), 29-47.

21. Govt&Resident_Fdls (2013) 'Boracay Island Coding Summary by Node Report' in V. Maguigad (ed.) *PhD Candidacy Fieldwork.* Townsville: James Cook University.

22. Govt_Jka (2013) 'Boracay Island Coding Summary by Node Report' in V. Maguigad (ed.) *PhD Candidacy Fieldwork.* Townsville: James Cook University.

23. Govt_Rc (2013) 'Boracay Island Coding Summary by Node Report' in V. Maguigad (ed.) *PhD Candidacy Fieldwork.* Townsville: James Cook University.

24. Govt_Re (2013) 'Boracay Island Coding Summary by Node Report' in V. Maguigad (ed.) *PhD Candidacy Fieldwork.* Townsville: James Cook University.

25. Govt_Rl (2014) 'Boracay Island Coding Summary by Node Report' in V. Maguigad (ed.) *PhD Candidacy Fieldwork.* Townsville: James Cook University.

26. Grove, K., & J. Pugh (2015) 'Assemblage Thinking and Participatory Development: Potentiality, Ethics, Biopolitics', *Geography Compass,* 9(1), 1-13.

27. Grydehøj, A. (2014) 'Constructing a Centre on the Periphery: Urbanization and Urban Design in the Island City of Nuuk, Greenland', *Island Studies Journal*, 9(2), 205-222.

28. Grydehøj, A., X.B. Pinya, G. Cooke, N. Doratli, A. Elewa, I. Kelman, J. Pugh, L. Schick, & R. Swaminathan (2015) 'Returning from the Horizon: Introducing Urban Island Studies', *Urban Island Studies*, 1, 1-19.

29. Harmeling, S. (2011) *Global Climate Risk Index 2012.* Berlin: Germanwatch E.V.

30. Interaksyon (2014) 'New Construction Moratorium Policy up in Boracay',*Interaksyon*. Available at http://interaksyon. com/ article/99570/new-construction-moratorium-policy-up-in-boracay. Accessed 11 November 2014.

31. Macapagal-Arroyo, G. (2006) *Proclamation 1064 Classifying Boracay Island Situated in the Municipality of Malay, Province of Aklan into Forestland (Protection Purposes) and into Agricultural Land (Alienable and Disposable) Pursuant yo Presidential Decree No. 705 (Revised Forestry Reform Code of the Philippines)*. Manila: Republic of the Philippines.

32. Maguigad, V. (2013) 'Tourism Planning in Archipelagic Philippines: A Case Review', *Tourism Management Perspectives,* 7, 25-33.

33. Maguigad, V., D. King, & A. Cottrell (2015) 'Towards A Political Ecology Approach To Island Tourism Urban Planning And Climate Change Adaptation In The Philippines: A Methodological Exploration', *Urban Island Studies*, 1, 133-151.

34. Mimura, N., L. Nurse, R. Mclean, J. Agard, L. Briguglio, P. Lefale, R. Payet, & G. Sem (2007) 'Small Islands' in *Climate Change 2007: Impacts, Adaptation and Vulnerability. Contribution of Working Group II to the Fourth Assessment Report of the Intergovernmental Panel on Climate Change.* Cambridge, UK & New York, NY: Cambridge University Press.

35. NGO_LT (2013) 'Boracay Island Coding Summary by Node Report' in V. Maguigad (ed.) *PhD Candidacy Fieldwork.* Townsville: James Cook University.

36. Nicholls, R.J., P.P. Wong, V. Burkett, J. Codignotto, J. Hay, R. Mclean, S. Ragoonaden, & C.D. Woodroffe (2007) 'Coastal Systems and Low-Lying Areas' in*Climate Change 2007: Impacts, Adaptation and Vulnerability. Contribution of Working Group II to the Fourth Assessment Report of the Intergovernmental Panel on Climate Change.* Cambridge, UK & New York, NY: Cambridge University Press.

37. NSCB-PSA (2014) *Philippine Standard Geographic Code: Malay, Aklan.* Manila: Philippine Statistics Authority - National Statistical Coordination Board.

38. Nurse, L.A., R.F. Mclean, J. Agard, L.P. Briguglio, V. Duvat-Magnan, N. Pelesikoti, E. Tompkins, & A. Webb (2014) 'Small Islands' in *Climate Change 2014: Impacts, Adaptation, And Vulnerability. Part B: Regional Aspects. Contribution of Working Group II to the Fifth Assessment Report of the Intergovernmental Panel on Climate Change.* Cambridge, UK & New York, NY: Cambridge University Press.

39. Nvivo (2014) 'How are Cluster Analysis Diagrams Generated?', *Nvivo.* Doncaster, AU: QSR International.

40. Ong, L.T.J., D. Storey, & J. Minnery (2011) 'Beyond the Beach: Balancing Environmental and Socio-Cultural Sustainability in Boracay, the Philippines', *Tourism Geographies,* 13(4), 549-569.

41. PCCC (2014*) Supplemental Guidelines on Mainstreaming Climate and Disaster Risks in the Comprehensive Land Use Plan.* Manila: Philippine Climate Change Commission.

42. Picornell, M. (2014) 'Insular Identity and Urban Contexts: Representations of the Local in the Construction of an Image of Palma (Mallorca, Balearic Islands)',*Island Studies Journal*, 9(2), 223-238.

43. Pigou-Dennis, E., & A. Grydehøj (2014) 'Accidental and Ideal Island Cities: Islanding Processes and Urban Design in Belize City and the Urban Archipelagos of Europe', *Island Studies Journal*, 9(2), 259-276.

44. Pugh, J. (2013) 'Speaking without Voice: Participatory Planning, Acknowledgement, and Latent Subjectivity in Barbados', *Annals of the Association of American Geographers*, 103(5), 1266-1281.

45. Resident&Tourismbiz_JDT&JCT (2013) 'Boracay Island Coding Summary by Node Report' in V. Maguigad (ed.) *PhD Candidacy Fieldwork.* Townsville: James Cook University.

46. Resident&Tourismbiz_MW (2013) 'Boracay Island Coding Summary by Node Report' in V. Maguigad (ed.) *PhD Candidacy Fieldwork.* Townsville: James Cook University.

47. Salvacion, A.R., & D.B. Magcale-Macandog (2015) 'Spatial Analysis of Human Population Distribution and Growth in Marinduque Island, Philippines', *Journal of Marine and Island Cultures*, 4(1), 27-33.

48. Shatkin, G. (2008) 'The City and the Bottom Line: Urban Megaprojects and thePrivatization of Planning in Southeast Asia', *Environment and Planning,* 40(2), 383-401.

49. Shatkin, G. (2000) 'Obstacles to Empowerment: Local Politics and Civil Society in Metropolitan Manila, the Philippines', *Urban Studies,* 37(12), 2357-2375.

50. Smith, R.A., J.C. Henderson, V. Chong, C. Tay, & Y. Jingwen (2011) 'The Development and Management of Beach Resorts: Boracay Island, the Philippines',*Asia Pacific Journal of Tourism Research,* 16(2), 229-245.

51. Thompson, M., R. Ellis, & A. Wildavsky (1990) *Cultural Theory.* Boulder: Westview.

52. Tourismbiz&Ngo_Ds&Pm (2013) 'Boracay Island Coding Summary by NodeReport' in V. Maguigad (ed.) *PhD Candidacy Fieldwork.* Townsville: James Cook University.

53. Tourismbiz_JC (2013) 'Boracay Island Coding Summary by Node Report' in V. Maguigad (ed.) *PhD Candidacy Fieldwork.* Townsville: James Cook University.

54. Tourismbiz_JCS (2013) 'Boracay Island Coding Summary by Node Report' in V. Maguigad (ed.) *PhD Candidacy Fieldwork.* Townsville: James Cook University.

55. Tourismbiz_LR (2013) 'Boracay Island Coding Summary by Node Report' in V. Maguigad (ed.) *PhD Candidacy Fieldwork.* Townsville: James Cook University.

56. Tourismbiz_ML (2013) 'Boracay Island Coding Summary by Node Report' in V. Maguigad (ed.) *PhD Candidacy Fieldwork.* Townsville: James Cook University.

57. Tourismbiz_RA (2013) 'Boracay Island Coding Summary by Node Report' in V. Maguigad (ed.) *PhD Candidacy Fieldwork.* Townsville: James Cook University.

58. Trousdale, W.J. (1999) 'Governance in Context: Boracay Island, Philippines',*Annals Of Tourism Research,* 26(4), 840-867.

59. Trousdale, W.J. (1997) *Carrying Capacity Considerations: The Need for Managing Change in a Unique Tourism Destination (Boracay Island Philippines).* Vancouver: Ecoplan International.

60. Visitor_ACG (2014) 'Boracay Island Coding Summary by Node Report' in V. Maguigad (ed.) *PhD Candidacy Fieldwork.* Townsville: James Cook University.

61. Visitor_PJS&ADR (2013) 'Boracay Island Coding Summary by Node Report' in V. Maguigad (ed.) *PhD Candidacy Fieldwork.* Townsville: James Cook University.

62. Welsh, E. (2002) 'Dealing with Data: Using Nvivo in the Qualitative Data Analysis Process', *Forum Qualitative Sozialforschung,* 3(2), Art. 26.

3

# CULTURAL REALIGNMENT, ISLANDS AND THE INFLUENCE OF TOURISM

**Donald Macleod**

University of Glasgow

## ABSTRACT

This article introduces a new concept: 'cultural realignment', which embraces phenomena such as cultural representation, interpretation, stereotyping and branding. Cultural realignment is the intentional depiction or interpretation of a culture (or part of one) for a specific preconceived purpose. It relates directly to power, and there is a need for this broad concept to help comprehend processes in an era of increasing globalisation, the growth of cultural commodification and the proliferation of representations in media including the internet. A prime concern of the article is the way that cultural realignment impacts on the identities of the communities subject to the realignment. The main examples given relate to island communities and their representation by anthropologists, and to island tourist destinations that have been subject to various descriptions, physical transformations and commodification driven by the tourism industry. A case study is examined as an example in the Canary Islands, using original research material related to recent and longitudinal fieldwork.

## KEYWORDS

Cultural realignment, identity, islands, tourism, anthropology, Canary Islands, La Gomera


## 1. INTRODUCTION

This article investigates processes and impacts that include and involve creating or changing a representation of a cultural group and associated place. These processes are to be denoted by a new overarching term Cultural Realignment. It can relate to the marketing of images and branding of a group of people, dwelling place or cultural site; the promotion or reorganisation of tangible and intangible heritage; as well as the written description of such phenomena where the specific intention is to realign the subject matter. As such it is a new concept that embraces others such as cultural representation, cultural interpretation and cultural commodification. Island communities provide the focus of the article as they make appropriate examples and have featured prominently throughout anthropological writings, the central disciplinary approach of this paper; additionally, they play a prominent part in the history and development of tourism.

The primary concerns of this article are the actual processes of cultural realignment, the actors involved and the impacts of the processes on inhabitants of the places. It will also explore the qualities of islands, especially in relation to anthropology, identity and tourism. It is argued that the process of realignment is growing rapidly in occurrence, especially because of tourism, and has major impacts. This new concept is suggested to help answer the question: how can we improve our understanding of cultural change, especially concerning tourism and island communities? The case study provided shows some of the processes and results of tourism-driven cultural realignment, which is able to impact on all levels of society from nation-state to villages and their inhabitants.

Anthropology is especially suited to investigate these influences due to its focus on ethnographic fieldwork and its holistic bias, as well as its method of comparative analysis and it will provide the main disciplinary approach, reflecting the researcher's background. Moreover, anthropologists have had a strong history of work on islands. The newly emerging multi-disciplinary study of islands (eg Baldacchino and Milne, 2000) suggests their potential for utilisation as models for other types of locations, implying that the improved understanding of the specific influences of tourism on islands has importance far beyond their shores.

## 2. ISLANDS AND ANTHROPOLOGY

Islands are good to think with. This statement paraphrases Levi-Strauss (1969) who noted that 'animals are good to think with' in relation to the way that different human cultures have used animals as symbols and as means of classification, creating boundaries and identities in human society. In a similar manner, writers and scholars have used islands (both real and fictional) as symbols and models to represent places and communities where distinct and unusual events happen, or to explain circumscribed communities, seemingly isolated, relatively untouched by other cultures. This attribute of islands is a cultural construction and has appeared in literature and art in western societies as well as others including Japan and Indonesia. More pertinent for this article however, is the fact that the island image of a self-contained homogeneous unit has been very appealing to anthropologists seeking distinctive groups of people. Indeed, for many anthropologists, islands have been very good to think with. This tendency has altered in time with the broadening of research interests, the general challenge to essentialism and a greater awareness of communication and acculturation. However, other groups in society continue to represent islands in a way that does not reflect their actual cultural composition, and one such group is the tourism industry

Islands have featured prominently in anthropological work. Malinowski's experience in the Trobriand Islands (Malinowski, 1922) was instrumental in the paradigm shift towards ethnographic fieldwork and a powerful model, especially in British Anthropology (Kuper, 1989). His focus on small, seemingly insular, discrete communities inspired others and preceded many influential anthropological outputs based on island experiences (eg Bateson, 1939; Firth, 1936; Geertz, 1963, Mead, 1928). This might be partially explained by a desire to research a discrete community, along functionalist lines, perhaps an isolated community which retains its culture and traditions, or demonstrates a unique pattern of social organisation, cosmology, kinship patterns, economic activity, and so on. In this the island represents or contains a 'whole' culture, one that is physically and metaphorically bound and relatively isolated from other influences. This was certainly a motivating factor for some anthropologists, and it has been criticised as leading to deliberate ignoring or diminishing of the socio-cultural variety of the island community, as well as intercultural impacts from overseas such as colonialism, missionary work, trading exchange and eventually, tourism when writing about the community (cf Eriksen, 1993; Kuper, 1989: 34). Perhaps the earlier writers were realigning the fieldwork material into a readable product suited to their own worldview or a particular intellectual theory.

More recently, anthropologists have continued to research on islands, sometimes with similar aims to the early pioneers (Herzfeld, 1991; Palsson, 1991) but also, and contrastingly, sometimes to investigate the strong influences of external phenomena: acculturation, trading, colonialism, migration, and tourism (eg Boissevain, 1993, 2010; Cohen, 1982; Crick, 1994; Mintz, 1960; Pickering, 2010; Sutherland, 1986; Waldren, 1996). The island often represents (sometimes erroneously) the notion of cultural uniqueness, a level of inaccessibility, with water as a physical and cultural boundary: all are factors especially pertinent for the examination of identity.

# 3. ISLANDS, IDENTITY AND THE EXTERNAL MARKET

Islands, because of their discrete, physically clear boundaries have often served well to describe the parameters of individual groups of people, even though there may have been strong communication networks by sea or air: again, they are good to think with. Consequently they also serve to act as models of impact analysis, whereby specific groups encounter other groups through various processes of acculturation. For some they have served almost as laboratories, bound and observable, for medical experiments as well as for isolating sectors of society. They have also been seen as repositories and examples of cultural survivals in terms of practices, arts, social organisation or systems of belief. Islands have been seen as distinct by artists, writers, natural scientists, governments, islanders themselves and by some social scientists. Anthropologists researching aspects of identity on islands have included Waldren (1996) on insider-outsider relationships, Cohen (1982) on belonging, Galvan Tudela (1987) on festivals and religion, and Macdonald (1997) on cultural heritage

Despite the rich body of work described above, there has been little research by anthropologists into the market-oriented cultural realignment of island identity in terms of cultural heritage interpretation, and island branding, its actual process (actors and agency, structural relationships and power), as well as its impacts on local populations; although Selwyn (1996) provides a relevant collection on mythmaking in tourism. In recent years this type of representation has become a large part of island identity construction in terms of an image for consumption by outside parties. This may be combined with more local and specific image building and reinvention, such as new interpretations of cultural heritage (both tangible and intangible), all of which activities can be considered to be 'cultural realignment'. Islands are particularly vulnerable to this type of activity if they are tourist attractions because they need to distinguish themselves in the marketplace, and are likely to become commodified: increasingly, culture has become the differentiating factor, a product and an added attraction.

Other social scientists have worked in areas closely related to the cultural realignment of identity and place as oriented to an external market and these include the following: Urry (1990) who considered the way that tourists gaze at places, presuming the relevance of their home country experience in fashioning the way they see other places as tourists; MacCannell (1976) drawing attention to the iconic status of some assets which come to represent a country, acting as major attractions (eg the Eifel Tower); Dann (1996) reviewing tourist brochures, analysing their content and their subsequent appeal to a specific market segment; Hall (1999) examining the branding of place as part of the touristification of a destination; Di Domenico (2001) who looked at the specific example of Dundee, which was branded as 'City of Discovery' after the ship built there, somewhat in contrast to its reputation for 'jute, jam and journalism'; and Sheller (2003) who explored the ways Europe and the USA 'consumed' numerous aspects of the Caribbean. These writers deal with the way tourism has influenced the manner in which a place and occasionally its inhabitants have been represented to others, with cultural commodification and stereotyping being the closest phenomena to the cultural realignment theme of this article. However, there are insufficient writers looking at the broader impact of such realignment on the inhabitants of the destination. Macleod (2009) drew attention to the development of theme towns in Scotland, including Wigtown Book Town, and the obsession of people with the economic impact to the detriment of concern with socio-cultural impacts. In a thesis Press (2001) had suggested that the nascent 'Book Town' of Wigtown initially caused great disquiet among the inhabitants who thought that they and their town had no strong connections with the book trade: the new identity was not appropriate. These writings both deal with the agency of cultural realignment and its impact on the actual, individual inhabitants of the towns in question; they exemplify how the process can embrace some of the smallest collections of people, as well as the largest groups such as nations, especially where marketing promotion for tourism is concerned.

# 4. CULTURAL REALIGNMENT

The term 'cultural realignment' is a concept developed in order to describe numerous processes in which people deliberately, with particular intention, create, change, manipulate, or interpret a culture or aspects of a culture (both tangible and intangible) for a preconceived purpose; for example to market the culture, or translate it into their own personal framework in order to understand it. They seek to 'change something to a different position or state' in order to line it up with a set of objectives. It is a particularly appropriate term to describe activities observable in the study of tourism.

There is a strong link between cultural realignment and power in the sense of having the ability to portray a culture and publicise the realignment or more dramatically realign it with a view to changing the lives of the subject population through tourism development or national identity building. Furthermore, it has strong connections with other concepts and processes, aspects of which it can embrace as described below: interpretation, discourse, propaganda, commodification, marketing, cultural engineering and zoning. It can build on these concepts and offers a new overarching concept embracing certain shared aspects. Thus not all discourses fall into cultural realignment; but one that seeks to portray a culture in a particular way in order to promote it for tourism does so.

Cultural realignment occurs in the written description of a culture and its components, it relates to the work of many anthropologists, in the production of monographs for example, and the attempt to organise findings to fit a cognitive model that would be comprehended by a particular audience. This has resonances with the concept of 'writing culture' (Clifford and Marcus, 1986) as well as 'interpretation of culture' (Geertz, 1973). It also relates directly to the work of others with political or economic interests, for example governments, journalists and marketing agencies when organising, promoting and generally influencing aspects of a group of people or place with a definite aim and objective. I would argue that 'Orientalism' is an example of cultural realignment on a grand scale (see Said, 1978).

Furthermore, cultural realignment has similarities with the concept of a 'dominant discourse' (after Foucault) in the sense of accepted ways of talking and thinking about things promoted by an elite group; but discourse itself does not adequately describe the physical changing of material things such as buildings as heritage, or the marketing of a place which does not necessarily follow dominant representations. Similarly, the concept of 'hegemonic representation' (after Gramsci), which includes ideas promoted by the ruling class, has parallels with cultural realignment, but again does not always explain the realignment of a culture, or aspects of a culture, designed to please anticipated visitors, such as tourists, or the realignment promoted by a minority group.

One example of cultural realignment which relates to the above theories is nation building, in the sense of drawing together histories, events, folk and public heroes, symbols and so on in order to create a sense of nationality for the citizens; as well as developing an image to present to the external, international community, the island state of Singapore being a good example (see Anderson, 1983; Bhandari, 2011; McCrone et al, 1995).

Britain may be a relatively recent invention historically, but it is from the reconstruction of the past through stories, and symbols of island-living, of war and of empire that an idea of Britishness in the present is re-created (Guiberanu and Goldblat, 2000: 135).

However, more pertinent to this article are examples of cultural realignment that include the commodification of culture (see Greenwood, 1989; Mckean, 1989) whereby culture, including performance and handicrafts becomes an object of exchange on the market. An awareness of the value of certain aspects of culture has led to the realisation that it can become a saleable commodity. This is a common development worldwide where tourism takes place. Another example is seen with the heritage industry: historical events and artefacts are packaged in a way to please a specific audience, often ignoring scholarly interpretation and authentic analysis in order to appeal or entertain (Hewison, 1987; Lowenthal, 1998; Stanton, 1989). Advertising media, including tourism brochures, are another example, in which descriptions, symbols and images are all marketed to appeal to a specific group of potential visitors concentrating on traits recognised by market analysis as being of interest to the targeted market segments (see Dann, 1996).

Not only is national identity subject to manipulation for the attention of possible visitors, but so is local identity on the smallest of scales. For example, a village might remain presented, portrayed and marketed as 'a fishing village' by the national tourism agency despite the fact that tourism has led to the fishers' demise, through offering alternative employment, purchasing beaches for hotel development, or driving fish away (Macleod 2001). New crops may be planted to give the landscape the look of a 'desert island', such as palm trees, or Mediterranean olive groves cultivated to appeal to visitors' desires, as they have been on Malta (see Billiard, 2011; Theuma, 2006). A destination may be culturally misrepresented for tourism purposes and attract a type of tourist or activity that the local people are not in favour of, or may be in stark contrast to their general tourism planning scenario, as was the case with Mafia Island (Tanzania) and a proposed dance festival (Caplan, 2011) which was perceived as a threat to their customs and culture.

Cultural realignment can embrace the transformation of the built and natural environment to conform to the desire of the potential visitor (see Selwyn, 2004 on cultural appropriation). More prosaically, cultural realignment can lead to 'cultural engineering' (see Hannertz, 1992) resulting in physical manipulation for cultural goals: transforming fishing ports into marine leisure harbours and ferry terminals. The imposition of national objectives might lead to the transformation of a peripheral island and its physical appearance as well as economic profile, as tourism is privileged over fishing and agriculture (Macleod, 2002).

Cultural realignment becomes an aspect and manifestation of power that reveals the relationship between the agents of change and those subject to it. The island of Bali is an example: Picard (1995) explains that Bali risked becoming a 'living museum' after the encouragement of Dutch colonial powers. Agents of change such as governments (national and local) and business organisations are able to promote their ideas, advertise images, publish and act on them, imposing these concepts on others. It can lead to the actual suppression of culture, including languages, a specific part of history, or an ethnic group; or in contrast, it may lead to the exaggeration and emphasis of such traits (see Lowenthal, 1998). This may encourage the physical removal of unwanted people or material objects. The creation of some 'national parks' is an example, imposing a cultural concept on a place and its inhabitants, both human and non-human, and sometimes involving the removal of people from a zone (Macleod, 2001). More benign aspects include those intellectuals, such as anthropologists, who have ignored some facts and features of a social scenario in writing about cultures, minimising the impact and relevance of missionaries, colonial powers or tourists: see Kuper (1989) on Malinowski; also Clifford and Marcus (1986) on 'writing culture'.

Elite groups have been developing ideas and realigning culture for the purposes of the market, for example in the tourism sector: those employed by governments or large business organisations often have an education in marketing or business administration but not equivalent expertise in cultural heritage or the sociology of the community. Such groups have begun to realign the culture towards the demands of the actual market, such as potential tourists, enquiring into their desires using the internet as a research tool – a very new approach for this type of marketing – and place more emphasis on a demand-led and specifically nuanced image of the destination (Fedele et al, 2011).

Again, using the internet, a more democratic opportunity has developed for people to circulate widely their own opinions and image of a destination. This may be done by visitors as well as by inhabitants of a place. Moreover, a stronger awareness of heritage together with opportunities to promote it as a commodity (as well as for political objectives) has led the local populace to generate their own views on local heritage (Macleod, 2010). Cultural realignment, therefore, has become a possibility for a growing and diverse number of people, with varied repercussions for the actual population objectified.

## 5. ISLANDS AND TOURISM

Islands are particularly popular with tourists. This is related to the countries producing tourists, notably those in Northern Europe, where a search for sun, sea and sand has led to journeys to islands in the Mediterranean Sea, Atlantic Ocean, Caribbean Sea and latterly the Pacific Ocean. A body of literature has been produced in the UK alone celebrating islands or using them as the scene of activity (Robinson Crusoe, Treasure Island, Lord of the Flies, The Magus). The glamorous image of the island paradise is promoted by the film industry (Dr No, Thunderball, The Beach, Fantasy Island), as well as being a staple feature of the advertising industry. All these media have fed into the popular public image of islands informing holidaymakers in Europe and North America, especially those indulging in cruise tourism visiting islands. As these are two of the most prominent regions producing tourists worldwide it is fair to say that the interest in islands is a dominant factor for the global tourism industry which accounted for around 940 million international travellers in 2010 spending US$ 919 billion (UNWTO, 2011).

The impact of tourism on islands has been profound. It accounts for a very large proportion of the Gross Domestic Product for many (eg the Greek Islands, the Caribbean Islands, the Canary Islands, the Balaeric Islands, the Azores, the Maldives, the Seychelles, and the Hawaiian Islands). Tourism has influences on the local population in terms of cultural activity, work patterns (displacement), wealth and its distribution, property-values and style, land-use, attitudes to environment, behaviour (demonstration effect), kinship patterns, and gender roles with changing opportunities, wealth, jobs and freedom (see Apostolopoulos et al, 1996; Burns, 1999; Smith, 1989). Inevitably the type of tourism is highly relevant to the type of impact that the tourists have: alternative tourists such as backpackers who stay with local families will have different impacts from package tourists staying in all-inclusive hotels (see Macleod, 2004). Many of the above factors will impact on local, regional, professional, and personal identities.

In its broadest sense, the tourism industry (which includes tour operators, hotels, travel and transport companies, government tourist boards, and marketing departments) is especially aware of the value of tourism to the destination site and the need to market the product adequately.

Acknowledging that guidebooks are aimed at tourists and want to meet their expectations, narratives, image and identity elements respond to tourist needs, reflecting to a greater or lesser extent new trends, interests and established images. (Marine-Roig, 2011: 122).

Media such as tourism brochures, internet sites, cinema and advertising, will attempt to appeal to the known desires of the potential tourist, promoting the popular image of islands in a generic fashion, oblivious to specific distinctions of the location, depending on the segment audience they seek to entice. This leads to stereotyping and bland generalisations, as with the opening statement on the Dominican Republic in a Thomson brochure: 'Fabulous beaches and plenty of Caribbean zest – this island does paradise escapes with aplomb' (Thomson, 2011).

Islands are often perceived as single, undifferentiated homogenous units in terms of destination site marketing, and are promoted as such. This perception of them as a 'whole' and unified culture is similar to some anthropological realignments of the idealised isolated insular community. Nevertheless, an increasingly sophisticated market, a changing demography of potential tourists, and the heightened need to offer 'added value' to the product with a inique selling point mean that in the twenty-first century markets have brought 'culture' into the mix as a feature of the total island product.

Because of their small size, isolation and limited resources both human and natural, many islands can be vulnerable to dependency on one or two particular industries or products. Tourism is one of these industries, occasionally referred to as a type of 'mono-crop' (Selwyn, 2004). Scholarly work on the Caribbean especially indicates this trait of dependency (Besson and Olwig, 2005; Mintz, 1960; Pattullo, 1996; Sheller, 2003). The risk of exploitation by powerful players including ex-colonial masters, international business and other organisations is ever present. Equally, because of its large influence on the economy, many islands acquiesce to the demands of the tourism industry and its agents, including the marketing and imaging of the island, its cultural realignment towards the (anticipated) desires of the potential tourists. This tendency to realign island culture for marketing purposes is growing, boosted by the internet. The tourism industry as a driver of cultural realignment will have impacts on the identity of local inhabitants of these destinations through the creation of images and narratives in brochures, the advertising media and the commodification of culture.

## 6. LA GOMERA

This case study explores some of the issues examined earlieri . La Gomera is one of the smallest of the seven habitable islands that form the Canary Island archipelago (see Figures 1 and 2). It is circular, with a diameter of about 25 km, and has a population of some 22,000. Agriculture and fishing were traditionally the main pillars of the island economy, with many people working on banana plantations supplemented by produce grown privately on tiny smallholdings. A fishing industry developed in the 20th Century supplying tuna for mainland Spain; it reached its peak in the 1970s, but is diminishing rapidly. Tourism has become an increasingly important economic force since the 1980s, especially in the popular destinations such as Valle Gran Rey (VGR), the focus of this section. The port of Vueltas in VGR has seen professional fishing replaced in economic importance by maritime leisure activities including dolphin watching and diving. Tourism is an increasingly important industry in the region and is therefore able to influence the image and presentation of the island, its cultural realignment, especially through the activities of government agencies such as the Patronata Insular de La Gomera, tourism organisations such as TUI, and local private businesses.

The types of visitors to VGR have been predominantly German backpackers, 'alternative tourists' (Duysen, 1989; Macleod, 2004), who have had distinct and wide-ranging impacts due to their propensity to mix and communicate with the local population who host them in their apartments or spare rooms. This visitor pattern has changed since the 1990s with more families, older package tourists, activity oriented tourists and different nationalities joining the original type due to internet usage, European Union (EU) membership, more bookable accommodation and an increasing awareness of the destination.

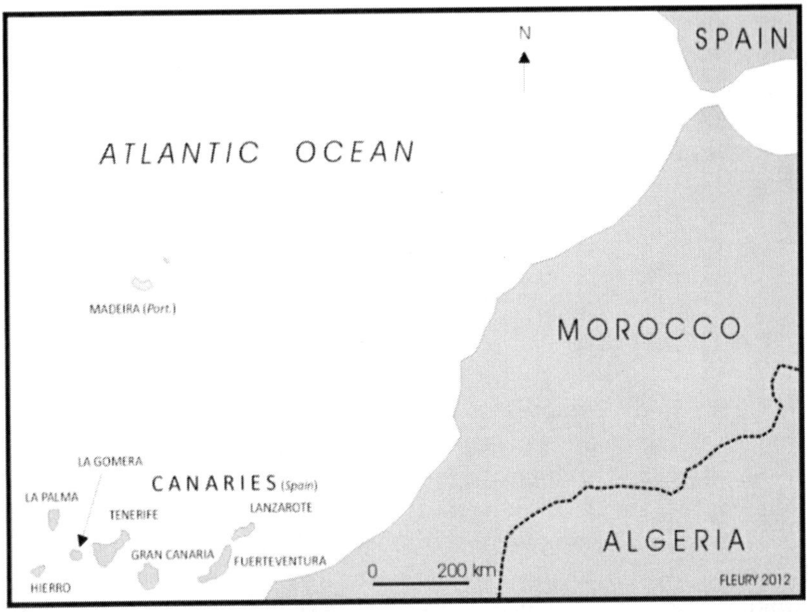


**Figure 1.** La Gomera, The Canary Islands and adjacent region (Christian Fleury, 2012)

# 7. GUANCHE HERITAGE AND THE INFLUENCE OF SPAIN

The original inhabitants of La Gomera were the Guanches, said to be related to the Berbers from North Africa (Hernandez-Hernandez, 1986). The Spanish colonised the island during the 15th Century and eventually subjugated the indigenous people. Vestiges of pre-Hispanic traits remain in the form of arts, crafts, and agricultural practices, through, for example, music, dance and the whistling language (silbo) (see Carreno, 2004). The ancient arts are celebrated during the numerous fiestas every year, and historic traditional handicrafts provide products for a growing tourist market.

Many people encountered during my research in VGR had a strong sense of being different from the people of mainland Spain and, to explain this historically, some referred to Guarapo (directed by Santiago Rios and Teodoro Rios, 1987), a popular film that depicted life on La Gomera during the 1950s, in which the islanders attempt to escape from the grips of the Spanish administration (represented by the Guardia Civil) to a better economic future in Venezuela via illegal passage across the Atlantic Ocean in barcos fantasma (ghost boats). Moreover, the film alluded to the wealthy landlord and professional classes known as caciques, and the poor peasant, working class population: this division remained apparent in the 1990s, and my research was predominantly among the working class of the region.

Various aspects and figures from colonial history have been promoted by the island council (Cabildo) through the tourism agency: these include Christopher Columbus who used the island as a final point for reparation before departing across the Atlantic Ocean. La Gomera is often marketed as La Isla Columbina ('The Island of Columbus'), especially through tourist brochures produced by the island government agency Paronata Insular de Turismo de La Gomera, and commemorative events such as the annual Columbus celebration, which itself has an arguably elitist character, including yacht racing. Links with colonial Spain are additionally emphasised through reference to, and the preservation of, built heritage in the port capital of San Sebastian. However, this had minimal resonance with those people of VGR encountered in this research, who saw themselves as descendents of the Guanches, as is also apparent in other records (see Concepcion, 1989; Hernandez Hernandez, 1986).

# 8. THE PROMOTION OF LA GOMERA

As an archipelago, the Canary Islands became a recognised politically unified group only after conquest by Spain; beforehand the islands were separate independent political units governed by a system of kingship, in which communication between the islands was at a minimum (Hernandez Hernandez, 1986). Individual cultural traits remain idiosyncratic between the different islands: for example, La Gomera is home of the only whistling language in the archipelago. Unity has become increasingly vital following the conquest by Spain, and more recently with the islands as an autonomous region of Spain. This helps reduce weaknesses due to a relatively peripheral location and the associated economic resource problems of islands.

Despite their variety, the tourist industry historically promoted the Canary Islands as a single cultural entity until the early 1990s. Tenerife, with its long history of tourism, was highly promoted and has overshadowed its smaller sister island La Gomera with which it has strong political and social connections (many thousands of native Gomeros live in Tenerife). North European tour companies in particular established an image of Tenerife, making scant reference to La Gomera. With the increasing need to differentiate products in the market place, cultural factors have become very important to delineate one archipelago from another, and furthermore, at the level of the archipelago, to differentiate individual islands as well. Consequently, La Gomera has very recently been promoted for its cultural heritage, and money from the EU has been spent improving the built heritage in the capital port of San Sebastian. Colonial buildings have been completely renovated: those associated with Columbus (the Tower of the Count of Gomera, and a small museum dedicated to Columbus) greet the visitors as they disembark the ferry from Tenerife. At the port there is a tangible cultural realignment in progress: the physical preparation and popular promotion of La Gomera as a site of colonial Spanish heritage. One brochure produced by a tour company emphasises these features, but does not mention anything else about the culture of the islanders themselves:

The capital and main port, San Sebastian, with numerous restaurants, bars and shops is famed for its links with Christopher Columbus and the small town is riddled with memorabilia and plaques denoting the places that he visited or stayed, and a week-long Columbus Festival is held annually (Classic Collection, 2011: 50)

Millions of Euros have been spent on the Spanish colonial heritage and the wider tourism development infrastructure by government bodies but far smaller amounts spent on supporting artisanal fishing and farming or commemorating local fishing traditions, much to the frustration of some sectors of society (Guarapo, 1990). Examples of tourism development projects oriented to La Gomera include: the EcoPlan (1988); the PTE (2003) approved by the Canary Island Government to create zones relating to tourism development and delimiting land; and the plan entitled Estrategia de Tourismo Sostenible de La Gomera (2008-12) which has a focus on the World Heritage Site National Park 'Garajonay' and other key locations (see Bianchi 2004 on development plans).

## 9. HERITAGE COMMEMORATION AND CULTURAL REALIGNMENT

A fascinating development has been the recent erection of a statue of the Guanche hero Huatacuperche on Canary Islands Day (May 30th) 2007. He was a local man who challenged the leadership of the Spanish count, who had broken agreements, in the 15th Century. Huatacuperche led a short-lived rebellion in 1488, killing the count, which was roundly and cruelly put down by the Spaniards, according to historians (Concepcion, 1989). Indeed, Beatriz de Borbadilla, who was the surviving wife of the count was instrumental in much of the retributions delivered to the Gomeros, and pertinently for this topic, she was alleged to have had an amorous relationship with Columbus

A political grouping, Coalición Canarias, which controlled the local council of VGR for a period until 2011, sought to promote Huatacuperche as personifying the island's independent spirit. His image is currently boldly placed on the front of tourism brochures and may be regarded as strengthening the island's unique selling point (USP). This statue, some four metres high, has been placed in a prominent position overlooking a beach in front of the biggest hotel in VGR. Commemoration of the hero Huatacuperche, through a statue and the reproduction of his image on posters, pamphlets and brochures, illustrates – indirectly - the political scenario in the Canary Islands archipelago, wherein a coalition party promotes its own interests and those of this autonomous region of Spain (although this is not part of a move towards political independence). It has created an emblematic symbol in the figure of a heroic rebel, a son of La Gomera who is quickly becoming an iconic representation of qualities attractive to residents, and simultaneously becoming a beacon and iconic reference for visitors, similar to the Statue of Liberty in the USA. This is a tangible example of cultural realignment, involving history and built heritage, and supplies us with an interesting contrast to the other heavily promoted icon, Columbus.

Research undertaken by this author intermittently over a period since the early 1990s, shows that many of the alternative tourists visiting the island have not been attracted to the island because of its colonial heritage, but has shown that they do have an interest in the artisanal heritage related to agriculture and fishing, as well as a strong interest in the natural heritage manifest in the World Heritage Site (the National Park Garajonay, a temperate rain-forest) and the marine ecology exemplified through dolphin watching. This represents a mismatch with the way the island's cultural heritage is promoted through publicity material and investment, demonstrating a dissonance between the cultural realignment of the island by the governing organisations and the desires of many tourists.

One example of a local celebration of heritage is the commemoration of an historic event when people tried to escape the poverty of the island in the 1950s. Many islanders embarked on a dangerous journey across the Atlantic Ocean to Venezuela, in various types of craft, and some left in a yacht called 'Telemaco', crossing successfully. In the mid-1990s a fishing boat belonging to a local family was placed on a prominent road roundabout in the small port of Vueltas, as a memorial, together with a descriptive metal sign secured to a boulder. This story, and the entire period of poverty and misery known as La Miseria is very well known and dear to many of the island inhabitants; indeed, it forms the backdrop to the 1987 film Guarapo, much recommended to this researcher. The memorial demonstrates the values, priorities and concerns of this close-knit community, which contrast with the celebration of Columbus in San Sebastian.

# 10. REPERCUSSIONS OF REALIGNMENT

Through their experience of tourism, Gomeros have been introduced to other ethnicities, especially from Northern Europe; this has given them a direct insight into the culture and behaviour of others, as well as placing their own culture into a stronger comparative light. Their heightened awareness of their own individual culture has led in some areas to the increased commodification of activities that they can profit from (cultural heritage: pre-Hispanic and colonial). However, with commercialisation comes a certain loss of authenticity and meaning, as for example with the Fiesta Del Carmen, which celebrates the saint and protector of people at sea. This fiesta once took place over two days in 1991, dominated by the procession around the village boundary of the statue of the Virgin del Carmen, followed by the Catholic mass. By 2009 the fiesta was running for five days and nights, the majority of which were devoted to dancing and drinking with a fair held at the port harbour: religious celebration and local identity marking have become overshadowed by commercial interests.

Some people have complained to me about the loss of their culture in the sense of the commercialisation of fiestas such as Del Carmen described above, and San Marco during which, in the past, locals would build numerous bonfires in celebration. Furthermore, they complain about the difficulty of finding a quiet place on beaches which were once empty and are now overrun by tourists; the intensity of competitiveness between people in their working lives; and the lack of time to converse with friends. Although it must be added that there has been a reinvigorated interest in some aspects of cultural heritage among local people, regenerating traditional dancing and music, as well as the whistling language and interest in historic places and figures such as Huatacuperche. This may be thought of as a form of cultural involution (cf McKean, 1989).

Fishermen have been especially vocal about the dominance of recreational interests over fishing in the re-organisation and massive enlargement of the harbour, ongoing in 2011, which has come to favour leisure craft and ferries over fishing boats. One of my research informants, an elderly fisherman, furiously challenged the reorganisation of mooring places in the harbour bemoaning the fact that fishermen who grew up and helped develop the village and harbour were now being moved aside to make way for leisure craft. Moreover, a huge new extended harbour doubles the size of the original in order to contain a car-ferry. These are all repercussions, direct and indirect, of the cultural realignment of La Gomera along tourist destination lines.

# 11. CONCLUSION

Islands are excellent examples that demonstrate where cultural realignment has been happening, and continues to happen; whether driven by intellectuals such as anthropologists, writers and artists, or powerful groups such as governments and most pertinent to this article, the tourism industry. As such they are models for examination and provide comparative material for different types of geographical regions, nations and cities. Moreover, the use of islands forces us to reflect on why they have been so popularly employed and what insular attributes are so appealing.

Examples given have shown how cultural realignment may be driven by international organisations, national governments and local councils, as well as (more rarely) grassroots organisations. They demonstrate the far-reaching and major consequences of such processes on people's lives, changing livelihoods, environments and identities. Rapid globalisation and the growing pervasiveness of the internet impact on marketing and the democratic ability of people from all walks of life to promote personal, business and community interests. This has led the means of representation, interpretation and communication to transform quickly, enabling more groups and individuals to become active agents in the process of cultural realignment, with different degrees of influence depending on power. Nevertheless, we remain aware that cultures have always changed, albeit at different paces and in many ways: but their deliberate alignment towards an external market, especially for tourism, is new for many.

Cultural realignment is a new concept which has helped us to understand and link the various processes involved in cultural change, especially those involving images, representations and interpretations of islands through tourism, and by extension, other places. It embraces numerous concepts and processes including cultural representation, commodification, interpretation, stereotyping and branding; it has a distinct element of intentional agency with a predetermined objective, and can be manifest in physical heritage transformation as well as written and spoken forms

This article draws attention to the way that individuals and groups have sought to develop, interpret or market aspects of a culture with a view to influencing others, usually external to the culture. Those others may be potential tourists, or a group of intellectuals, or even the international political community. So they realign a collection of experiences, facts, or descriptions and orient them into a discourse palatable to a preconceived audience. They may even be realigning buildings, monuments or entire areas (different types of 'texts') for similar reasons. These processes are growing in occurrence, especially with the tourism industry, itself expanding and moving into new territory, encouraging change. There is a necessity to come to terms with these developments, the processes and their repercussions. This article seeks to place research findings and other work in a new light, helping scholars to comprehend the huge cultural manipulation that tourism encourages. It is hoped that this stimulates further research and discussion on the impacts and the processes of cultural realignment, as well as the importance of islands as the sites of social processes.

## END NOTE

[i] Research on La Gomera has been undertaken intermittently by the author since the early 1990s, beginning with a 12-month intensive anthropological fieldwork project, based on participant observation. The focus was on a fishing community in Vueltas, and foreign settlers and tourists based in the municipality of Valle Gran Rey. Occasional returns followed for short periods to update on changes, and in 2009 a six-week period of research into sustainable tourism development in the same area involved semistructured interviews and observation. The article includes original, unpublished findings from recent and earlier research periods. I am grateful to the Pirie Reid Scholarship Fund for supporting doctoral research 1989-91, and to the Carnegie Trust for funding fieldwork in 2009.

## BIBLIOGRAPHY

1. Anderson, B (1983) Imagined Communities: Reflections on the Origins and Spread of Nationalism, London: Verso

2. Apostolopoulos, Y, Leivadi, S, and Yiannikis, A (eds) (1996) The Sociology of Tourism: Theoretical and Empirical Investigations, London: Routledge

3. Baldacchino, G, and D. Milne, D (eds) (2000) Lessons from the Political economy of Small Islands: The Resourcefulness of Jurisdiction, London: Macmillan

4. Bateson, G. (1939) Naven, Cambridge: Cambridge University Press

5. Besson, J. and Olwig, K.F (eds) (2005) Caribbean Narratives of Belonging: Fields of Relations, Sites of Identity, Oxford: Macmillan

6. Bhandari, K. (2011) 'Recreating heritage in the southwest of Scotland', Current Issues of Tourism 2011 ni: 1-15

7. Bianchi, R.V (2004) 'Tourism restructuring and the politics of sustainability: A critical view from the European periphery (The Canary Islands)', Journal of Sustainable Tourism v12 n6: 495-529.

8. Billiard, E (2011) 'Tourism and National Identity in Malta', unpublished paper given at Second Island Dynamics Conference (Valletta, Malta, May 2011)

9. Boissevain, J (1993) Saints and Fireworks: religion and Politics in Rural Malta (3rd edition), Malta: Progress Press

10. (2010) 'Tourists, Developers and Civil Society: On the commodification of Malta's Landscape', In Scott, J and Selwyn, T (eds) Thinking Through Tourism, Oxford: Berg: 93-116

11. Burns, P.M (1999) An Introduction to Tourism and Anthropology, London: Routledge

12. Caplan, P (2011) 'A 'Clash of Civilizations' on Mafia Island? The story of a dance festival', Anthropology Today v27 n2: 18-21

13. Carreno, R.T (2004) 'El siblo gomero', in Canarias: La Gran Enciclopedia de la cultura, Teneriffe: Centro de la Cultura Popular Canaria: 56-59

14. Classic Collection (2011) Classic Collection Holidays Brochure, Winter Sun 2011/12 (West Sussex)

15. Clifford, J and Marcus, M.E (eds) (1986) Writing Culture: The poetics and the politics of ethnography, Berkeley: University of California Press

16. Crick, M (1994) Resplendent Sites, Discordant Voices: Sri Lanka and International tourism, Chur: Harwood Academic Publishers

17. Cohen, A.P (ed) (1982) Belonging: Identity and Social Organisation in British Rural Cultures, Manchester: Manchester University Press

18. Concepcion, J (1989) The Guanche Survivors and their Descendants, Tenerife: Concepcion

19. Dann, G (1996) 'The People of the Tourist Brochures', in Selwyn, T (ed) The Tourist Image: Myths and Myth Making in Tourism, Chichester: John Wiley and Sons: 61-81:

20. Di Domenico, M (2001) 'Re-imagining the City: Heritage, Tourism Strategies for Regeneration in Dundee' in Di Domenico, C, Law, A, Skinner, J, and Smith, M (eds) Boundaries and Identities: Nation, Politics and Culture in Scotland, Dundee: University of Abertay Press: 191-214

21. Duysens, B (1989) 'Turismo, Ocio y cultura Juvenil: El caso del turismo de mochila en La Gomera' ('Tourism, Recreation and Youth Culture') ERES (Antropologia) (Tenerife: Museo Arquelogico) v1 n2: 115-26

22. Ecoplan (1988) Ecoplan para la isla de La Gomera. Monografias de la Dirección General del Medio Ambiente, ('Ecoplan for Gomera island') Madrid: Ministerio de Obras Públicos y Urbanismo: 282-286

23. Eriksen, T. H (1993) 'Do cultural islands exist?, Social Anthropology n1: 1-16

24. Fedele, S, Marchori, E, Inversini, A, and Cantoni, L (2011) 'Understanding the online reputation of tourism destinations: The case of the Republic of Malta', unpublished paper given at Second Island Dynamics Conference (Valletta, Malta, May 2011)

25. Firth, R (1936) We The Tikopia: A Sociological Study of Kinship in Primitive Polynesia, London: Allen and Unwin

26. Galvan-Tudela, A (1987) Las Fiestas Populares Canarias ('Popular Festivals in the Canaries'), Tenerife: Ediciones Canarias

27. Geertz, C (1963) Pedlars and Princes, Chicago: Chicago University Press

28. (1973) The Interpretation of Cultures, New York: Basic Books

29. Giubernau, M and Goldblat, D (2000) 'Identity and Nation', in Woodward, K (ed) Questioning Identity: Gender, Class, Nation; London: Routledge: 115-155:

30. Greenwood, D (1989) 'Culture by the Pound: An anthropological perspective on tourism as cultural commodification', in Smith, V. (ed) Hosts and Guests: The Anthropology of Tourism (2nd Edition), Philadelphia: University of Pennsylvania Press: 171-186

31. Guarapo (1990) Guarapo: Associación Cultural y Ecologista v1 n1

32. Hall, D (1999) 'Destination branding, niche marketing and national image projection in central and eastern Europe', Journal of Vacation Marketing v5 n3: 227-237

33. Hannerz, U (1992) Cultural Complexity: Studies in the social organisation of meaning, New York: Columbia University Press.

34. Hernandez Hernandez, P (1986) Natura y Cultura de las Islas Canarias ('The Nature and Culture of the Canary Islands'), Tenerife: Litografia P.H. G. Romera S.A.

35. Herzfeld, M (1991) A Place In History: Social and monumental time in a Cretan Town, Princeton: Princeton University Press

36. Hewison, R (1987) The Heritage Industry: Britain in a Climate of Decline, London: Methuen

37. Kuper, A (1989) Anthropology and Anthropologists, London: Routledge

38. Levi-Strauss, C (1969) The Savage Mind, Chicago: University of Chicago Press

39. Lowenthal, D (1998) The Heritage Crusade and the Spoils of History, Cambridge: Cambridge University Press

40. MacCannell, D (1976) The Tourist: A new theory of the leisure class, London: Methuen

41. Macdonald S (1997) Reimagining Culture: Histories, Identities and the Gaelic Renaissance, Oxford: Berg

42. Macleod, D.V.L (2001) 'Parks or People? National Parks and the Dominican Republic', Progress in Development Studies v1 n3: 221-235

43. (2002) 'Disappearing Culture? Globalization and a Canary Island Fishing Settlement', History and Anthropology v13 n1: 53-67

44. (2004) Tourism, Globalisation and Cultural Change: An Island Community Perspective, Cleveland: Channel View Publications

45. (2009) 'Scottish Theme Towns: have new identities enhanced development?' Journal of Tourism and Cultural Change v7 n2: 133-145

46. (2010) 'Power, Culture and the Production of Heritage', in Macleod, D.V.L and Carrier. J.G (eds) Tourism, Power and Culture: Anthropological Insights, Bristol: Channel View Publications: 64-89

47. Malinowski, B (1922) Argonauts of the Western Pacific, London: Routledge and Kegan Paul

48. Marine-Roig, E (2011) 'The image and identity of the Catalan coast as a tourist destination in twentieth-century tourist guidebooks', Journal of Tourism and Cultural Change Volume v9 n2: 118-139

49. McCrone, D, Morris, A and Keily, R (1995) Scotland the Brand: The Making of Scottish Heritage, Edinburgh: Polygon

50. McKean, P.F (1989) 'Towards a theoretical analysis of tourism: Economic dualism and cultural involution in Bali', in Smith, V Hosts and Guests: The Anthropology of Tourism (2nd edition), Philadelphia: University of Pennsylvania Press: 119-138

51. Mead, M (1928) Coming of Age in Samoa, New York: Mentor Books

52. Mintz, S (1960) Worker in the Cane: A Puerto Rican Life History, New Haven: Yale University Press

53. Palsson, G (1991) Coastal Economies, Cultrual Accounts: Human ecology and Icelandic discourse, Manchester: Manchester University Press

54. Pattullo, P (1996) Last Resorts: The Costs of Tourism in the Caribbean, London: Cassell

55. Picard, M (1995) 'Cultural Heritage and Tourist Capital: Cultural Tourism in Bali' in Lanfant, M, Alcock, J.B, and Bruner, E.M (eds) International Tourism: Identity and Change. London: Sage Publishers: 44-66

56. Pickering, L (2010) 'Past Imperfect: displacing Hawaiians as hosts in a 'drop out' community in Hawai'i in Bonisch-Brednich, B and Trundle, C (eds) Local Lives: migration and the politics of place, Farnham: Ashgate Publishers: 45-63

57. Press, S.R (2001) Reading Wigtown: Place identities in a Scottish Book Town, (unpublished) MA Thesis, University of Oregon, Eugene

58. PTE (2003) Plan Territorial Especial de Desarollo Touristico de la Isla de La Gomera, Tenerife: Canary Island Government

59. Said, E.W (1978) Orientalism: Western conceptions of the Orient, London: Routledge & Kegan Paul Ltd

60. Selwyn, T (ed.) (1996) The Tourist Image: Myths and Myth Making in Tourism, Chichester: John Wiley and Sons

61. (2004) 'Privatising the Mediterranean Coastline', In Boissevain, J and Selwyn, T (eds) Contesting the Foreshore: Tourism: society and

politics on the coast, Amsterdam: Amsterdam University Press: 35-60

62. Sheller, M (2003) Consuming the Caribbean, London: Routledge
63. Smith, V (1989) Hosts and Guests: The Anthropology of Tourism (2nd Edition), Philadelphia: University of Pennsylvania Press
64. Stanton, M.E (1989) 'The Polynesian Cultural Centre: A Multiethnic Model of Seven Pacific Cultures', on Smith, V (ed) Hosts and Guests: The Anthropology of Tourism (2nd Edition), Philadelphia: University of Pennsylvania Press: 247-264
65. Sutherland, A (1986) Caye Caulker: economic success in a Belizean fishing village, London: Westview Press
66. Theuma, N (2006) 'Malta: Reimaging the Mediterranean Destination', in Hall, D, Smith, M.K and Maraiszwesia, B (eds) In Tourism and the New Europe: The Challenges and Opportunities of EU Enlargement, Wallingford: CABI: 213-223
67. Thomson (2011) Faraway Shores: Thomson Brochure, November 2011 – October 2012
68. Urry, J (1990) The Tourist Gaze: Leisure and travel in contemporary society, London: Sage
69. UNWTO (2011) World Tourism Organisation (United Nations), online at: www.unwto.org - accessed August 2011
70. Waldren, J (1996) Insiders and Outsiders: Paradise and Reality in Mallorca, Oxford: Berghahn Books

4

# FEEDING TWO MILLION RESIDENTS AND TEN MILLION TOURISTS

**Dirk Godenau, Juan Sebastián Nuez Yánez**

University of La Laguna, Canary Islands

## ABSTRACT

The level of food self-sufficiency in the Canaries is low and decreasing. The growing demand for food, both due to demographic and touristic expansion and to the population's increased purchasing power, has not seen a corresponding increase in local food production. This paper details the factors behind the growing dependence on imported food, emphasising the role of insularity and the institutional framework of food production activity. Based on this diagnosis, the main courses of action are identified that could allow for the selective recovery of that portion of the local production that is intended for the internal market

## KEYWORDS

Islandness, food sufficiency, vulnerability, Canary Islands

# 1. ISLANDNESS AND FOOD SUFFICIENCY

Food can be supplied in any territory through multiple combinations of local production and importation. The local supply is affected by the availability of natural resources (water, soil, climate, etc.) and the characteristics of the production system (property structure, capitalisation, technology, etc.). The demand side is conditioned by the size of the population and its purchasing power. High demographic densities, in turn, can reduce the availability of arable land and alter production costs as compared to other locations with more available land. Greater degrees of economic development tend to be reflected in a shopping cart that is more diversified and less dependent on the seasonality of local production and its range of foods. In this sense, the process of economic development promotes the openness of local systems and the exchange of foods through trade.

Within this general context, what is unique about the food supply under conditions of insularity? The economic effects of insularity are the result of combining the (reduced) size and (dis)connection of insular systems. The size, measured using economic (GDP), demographic (population) and geographic (size) variables, usually assumes that islands are small in comparison to mainland economic systems 1 . The higher degree of exposure exhibited in small systems stems from the effects of scale, of the islands' smaller average size. The reason is obvious: if the planet were to be regarded as a single country, it would, by definition, be fully self-sufficient. Large states such as China or the U.S. are expected to have high degrees of self-sufficiency. The same can be said for the diversity of the food they produce. A country with a large expanse is more likely to enjoy a higher diversity of ecosystems and natural resources, a diversity that on smaller scales, with few exceptions, tends to be less. As a consequence, when evaluating the openness of an economic system, it is useful to compare similarly sized economies. An insular economy may in fact be more closed than its mainland counterpart once the size variable is taken into consideration.

The second dimension of the effects of insularity involves the connectivity the island maintains with the outside. While 'classical' analyses of insularity rely on concepts such as 'remoteness' and 'periphery', weighing the distance from the mainland as a drawback (disconnection) and a cause of higher supply prices, more recent analyses highlight the connectivity provided by maritime routes (see King, 2009:63 on 'nodal islands'). An economic assessment of the distance separating islands from mainland economies requires considering several nuances. First, the per unit and per kilometre transport costs are, save for certain exceptions, lower for maritime than land transport due to the lower physical friction between the vehicle and its surroundings. This is also reflected in this activity's carbon footprint. For example, transporting 10 tonnes of cargo by ship from China results in 1.3 tonnes of CO2 emissions. Transporting that same quantity from Alicante to Hamburg in trucks would emit 1.56 tonnes. (Amman et al, 2012: online).

Second, beyond energy efficiency, the cost of transport also depends on the technologies employed. Specifically, many islands played a historic role as supply sites in the expansion of intercontinental transport. Nowadays, the use of large container ships allows for continents to be linked directly without the need for intermediate sites, reinforcing the peripheral nature of insular economies as places with reduced market sizes. As a result, the effects of scale when supplying islands from the outside is evidenced in the smaller sizes and frequencies of the means utilised, generating higher costs. Third, the reference distance is not necessarily the nearest mainland coast. If the nearest site cannot offer a range of products or an appealing infrastructure, it may not present any supply possibilities. Relevant to this question are the institutional memberships of the islands in States or integrated economic blocks, since borders condition trading possibilities and costs (customs and non-customs duties, the extent to which States encourage trade through port infrastructure, etc.).

In this sense, institutional economics (North, 1990) recommend taking into account the context (embeddedness) and history (path dependency) of the systems analysed. When considering the time aspect, one should recall the insertion of many islands in maritime networks by European powers (eg Malta, the Canary Islands, the Azores and Madeira; Baldacchino, 2004). The overseas expansion of economic ties gave these islands a geostrategic value that had repercussions on the economic specialisation and their openness (plantation islands; Warrington and Milne, 2007). This international specialisation, in turn, conditions the population of these islands by immigration with links to this selective globalisation through commercial ties. The local resource base is not interpreted in these contexts exclusively as a source of food for the resident population, because these resources can comprise comparative advantages in a context of international specialisation. For example, the comparisons by Gough et al. (2010: 5) among the islands of Anuta, Tikopia, Bellona and Ontong Java underscore how "by comparison Bellona and Ontong Java have changed rather faster and more comprehensively than Anuta and Tikopia, where commerce is largely absent and links with markets sporadic at best. Quite simply, geography matters." When considering the spatial aspect, the geographic position of the island being analysed affects the natural conditions under which food is produced (tropical or subtropical locations, etc.); Greenland is not the same as Cuba. Different climates, geographies and soils require the adaptation of local food systems, and many traditional food patterns in island areas are related to this adaptation. This context must also include institutional integration. Islands can be independent States, parts of States or of other economically relevant supranational entities (one example being the European Union). This integration affects the competencies of local policies. As concerns food and agriculture, an island's institutional status determines how much sovereignty it enjoys to craft and implement its own policies.

Of interest is how developed countries have used their sovereignty to maintain and even increase their level of self-sufficiency despite, in some cases, not having comparative advantages in terms of domestically produced foods. What are the arguments for justifying the objective of reaching or maintaining a minimum level of self-sufficiency? Food satisfies an inescapable physiological need: we need to eat. This characteristic transforms food into an individual and collective concern and gives it a strategic value. All communities are interested, insofar as they are able, in ensuring its future supply. As a result, it should come as no surprise that economic policies typically aim to safeguard the domestic food industry. The instruments for this defence have varied over time and according to the power of the communities involved. Another reason for valuing food products rests on their quality: a balanced range of products that is accessible by the entire population is supremely important to the health and performance of a society. Obviously this is affected not only by supply factors, since eating unhealthy foods can also be explained by the voluntary behaviour and preferences of the population, but the lack of a balanced supply undoubtedly has a disproportionate effect on this behaviour

The strategic interest of food security is particularly relevant in the context of the higher vulnerability of small insular economies (United Nations, 1998). This vulnerability is exacerbated by the impact of climate change, particularly in some cases, such as those of the Pacific islands (FAO, 2008). Climate change is related to food supply both through its direct impact on local production and indirectly through imports. The goal of food security, defined as a situation in which "all people, at all times, have physical, social and economic access to sufficient, safe and nutritious food that meets their dietary needs and food preferences for an active and healthy life" (FAO, 2002; Chapman,

2009: 11), places the emphasis on the goal of making the local system as resilient as possible to potential adverse events that could impact the food supply.

In light of the considerable importance of imported food in the supply chain of many islands, most of the recommendations made to achieve food security involve recovering and diversifying local food production instead of devoting a large area of what little arable land is available to cultivation for export or to land development (FAO, 2008: 9; a recent example for the Pacific islands is Barnett, 2011: 232). There are also cases, however, of small islands whose populations feed themselves almost exclusively using locally grown food (Mertz et al., 2010). In these cases, food security could also be at risk if natural disasters were to suddenly and massively affect local production.

Given this interest in maintaining a minimum level of food self-sufficiency, the question arises regarding whether islands too can achieve this objective. The answer to this question depends on a combination of several interrelated factors. This paper analyses these factors in detail for the case of the Canary Islands. Before reviewing the evidence specific to the food market in the Canaries, a summary of its contextual characteristics is in order so as to relate the food and agricultural dimension to the islands' specific financial and demographic context.

## 2. THE CONTEXT OF THE CANARY ISLANDS

The Canary Islands are an archipelago consisting of seven islands and several rocky islands with a surface area of 7,477 square kilometres between 27 and 29 degrees north and 13 to 18 degrees west. Located 100 kilometres away from Africa and over 1,000 from Europe, the Cold Canary Current flows around the island before joining the Gulf Stream. The islands are under the constant influence of the Azores High, responsible for the archipelago's humidity-laden northwesterly winds (Figure 1). With its subtropical clime, the Current and the High result in very mild temperatures with slight differences between highs and lows, with an annual average on the coast of between 18 and 21 degrees, around 11 degrees between 500 and 1500 meters and 9 degrees above that altitude.

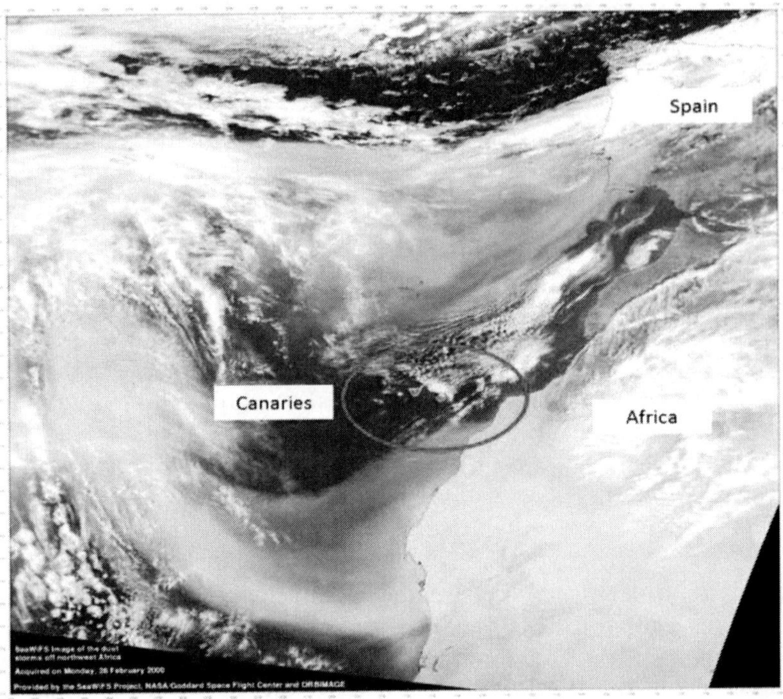


**Figure 1** Geographic position of the Canary Islands. (Source: NASA, 2000) II. The context of the Canary Islands

The geologically young islands are volcanic in origin and still active. This means that few rocks have turned into soil, resulting in only 16% of the islands' surface being arable. They are also very mountainous for their small surface area – the Teide volcano on the island of Tenerife reaches an altitude of 3,718 metres and is only 10 kilometres away from the coast – meaning there are very steep gradients. The small area and the gradient have conditioned the farming possibilities, with the non-existence of metal or coal mines limiting industrial activities.

These factors, plus the climate and geographic location, have determined the economic growth model used in the Canaries throughout their history, a model that is supported by three pillars: a food production economy based on exports, a service economy tied to international commerce, and a favourable institutional framework that has eliminated every roadblock to those outside ties (Macías, 2001). At one time, the export agriculture - replaced in the 1960s by tourism –

allowed for an influx of the currencies necessary for purchasing articles that were otherwise unobtainable. But alongside that agriculture there was always another intended to ensure the local population's supply of staple products. This aspect played a very important role despite its low level of capitalisation since, on the one hand, it served to keep salaries low, making exported articles more competitive, and on the other, when the exports struggled, it allowed the owning class to maintain its income level (Burriel de Orueta, 1982; Macías, 1981).

This connection between the two agricultural subsectors, however, began to unravel in the second half of the 19th Century, a schism that accelerated starting in 1900 and was complete after 1960. As improvements in international maritime transport led to lower freight costs, merchants in the Canaries opted to buy food from the outside, not only because it was cheaper, but because they could control the entire distribution process and thereby increase their profits. Agricultural production for local consumption thus started to languish, since it was unable to compete with the price of imported foods, being limited to those areas of the islands with little or no market access or practiced by the growing numbers of part-time farmers who use it to supplement incomes obtained from other economic activities (Sans, 1977 and 1981). Thus, the dishes typical of Canary Island cuisine, like papas arrugadas2 , are being prepared increasingly with products imported from Great Britain, Egypt or Israel and eaten with wines from the mainland (Nuez, 2008).

## 3. ESTIMATING FOOD SUFFICIENCY IN THE CANARY ISLANDS

### 3.1 Food demand

In recent decades, the size of the food market in the Canaries has grown both due to the expansion of the demographic base, which was 34% higher in 2010 than in 1995, and to the increased purchasing power of the population. Currently the Canary Islands archipelago is densely populated, with 2,115,049 inhabitants residing in an area of 7,447 square kilometres (1 January 2012), equivalent to a population density of 284 inhabitants per square kilometre. This situation is the result of high population growth in the 20th Century, particularly between the years of 1998 and 2008 and due primarily to immigration, with natural growth playing a minor role. To this resident population we must add the presence of approximately 10 million tourists every year who, when their average stay is translated into an average daily floating population, approximately amounts to a 10% increase (around 200,000 people).

The trend in per capita income for residents of the Canaries has shown a meagre real increase (at 2000 constant prices) of 10% between 1995 and 2010 (Figure 2). The economic expansion had allowed for higher levels in 2005, but the contraction in the GDP during the subsequent recession and the continuing population growth resulted in a substantial reduction over a short period of time. In 2011, the per capita GDP in the Canaries was 19,806 euros, well below the Spanish average of 23,271 euros and the European Union (EU-27) average of 25,200 euros.

| **Indicator** | 1995 | 2000 | 2005 | 2010 |
|---|---|---|---|---|
| Resident population | 100 | 109 | 124 | 134 |
| GDP (2000 prices) | 100 | 124 | 146 | 148 |
| pcGDP (2000 prices) | 100 | 113 | 118 | 110 |

**Figure 2.** Population, per capita income and GDP in the Canaries (1995=100). (Source: INE.)

As is usually the case, the process of economic development reduces the share of the amount spent on food products. While in 1981 food and drink purchases accounted for 33.1% of a family's expenses, that share had dropped to 17.6% by 2011 (Figure 3). The financial crisis that started in 2008 interrupted this structural trend, and the percentage of food expenditure is even seeing a modest recovery. The greater well-being of the population also changes the composition of the shopping basket (Figure 4), with more money being spent on animal products and processed foods at the expense of lower consumption of plant-based products. The exception in this area is fresh fruits and vegetables, whose higher demand-income elasticity allows them to maintain or even expand their position in the shopping basket.

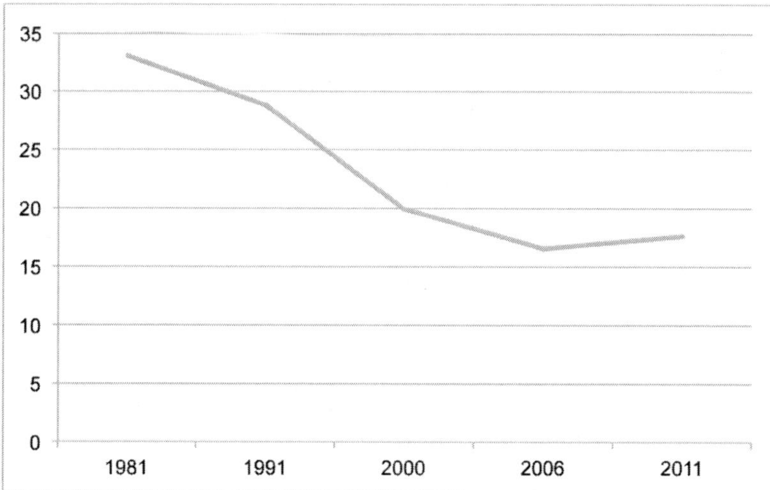


**Figure 3.** Percentage of family expenses allocated to food and non-alcoholic drinks. (Source: INE. Encuesta de Presupuestos Familiares [Family Budget Survey]).

| Food groups | 1988-1990 | | 2008-2010 | |
|---|---|---|---|---|
| | Volume | Value | Volume | Value |
| Meat | 5.4 | 17.4 | 6.0 | 19.5 |
| Fish | 2.1 | 7.4 | 2.8 | 9.5 |
| Milk and dairy products | 16.1 | 16.8 | 15.6 | 15.8 |
| Eggs | 2.1 | 2.0 | 1.1 | 1.2 |
| Bread and baked goods, cookies, cereals, chocolates | 8.9 | 12.1 | 7.3 | 11.7 |
| Rice, pasta, potatoes, vegetables | 14.1 | 6.9 | 6.6 | 3.4 |
| Oil | 3.4 | 3.5 | 2.2 | 2.3 |
| Fresh vegetables | 8.0 | 6.5 | 8.1 | 6.5 |
| Fresh fruits | 14.8 | 10.3 | 13.2 | 9.7 |
| Processed fruits and vegetables | 1.4 | 2.7 | 2.5 | 3.7 |
| Alcoholic drinks | 3.5 | 4.4 | 3.1 | 3.2 |
| Non-alcoholic drinks | 16.6 | 6.6 | 27.1 | 7.2 |
| Other | 3.7 | 3.4 | 4.4 | 6.2 |

**Figure 4.** Percentage distribution of the per capita expense (value) and consumption (volume) in Canary households by food group. (Source: Spain's Ministry of Agriculture, Food and the Environment.)

According to the statistics from the Food Consumption Panel of Spain's Ministry of Agriculture, Food and the Environment, the average annual amount spent by households in the Canaries on food and drink from 2008 to 2010 was 3.078 million current euros, with a volume of 1.5 million tonnes. This corresponds to a per capita volume of 765 kilograms a year and an annual budget of 1,555 euros. These data do not include consumption outside the home (HORECA: hotels, restaurant and catering) or the consumption of food not bought in markets. As a result, the food market volume in the Canaries measured as apparent consumption (production + imports – exports) will be noticeably higher3.

Changes in food demand in the Canaries involve a transition from a traditional food system to a modern one (Contreras and Gracia, 2005: 56) whose production is increasingly unrelated to the population, where access to food is primarily through the market mechanism with a distribution system centred around national and international chains and consumption patterns that are more and more homogeneous and independent of seasonal production. In this sense, changes in food consumption and purchasing habits in the Canaries have the same overall determinants as in developed countries: increasing numbers of wage earners with the corresponding effects on food consumption (more eating out, higher appreciation for the ease of use of food products); changes in family structures (smaller households, demographic aging, cultural diversification through immigration, reduction in and externalisation of house work); changes in the food supply (industrialisation of production, concentration of the distribution system, foods from distant places, specialisation of local foods in certain segments); and changes in attitudes, ideas and values (greater individualism, diversification of life styles).

The specific aspects of the situation in the Canaries are related to the effects of insularity and their reliance on tourism. The remoteness of the primary food sources, with food imported from the European mainland, and the fragmentation of the archipelago into seven insular markets of widely varying sizes condition the food supply through transportation and transformation costs. Reconstituted foods and drinks abound in the Canaries (cost savings by eliminating cheap ingredients like water before transportation and adding it back in later), as do preserved foods (eg frozen meat). Moreover, the so-called "double insularity" affects the structure of a regional market in terms of having products produced on all seven islands. Despite the complementarities among the insular products, stemming from the islands' different climatological and edaphological conditions, the inter-island transportation matrix for local foods is very underdeveloped.

The insular distribution structures, in turn, exhibit clear differences (Gutiérrez & Godenau, 2010), with the presence of large supermarket chains being more common in the islands with the largest market size (Gran Canaria, Tenerife), while in the smaller islands (El Hierro, La Gomera, La Palma), traditional commerce is more prevalent. The islands of Fuerteventura and Lanzarote, which cater more to tourists, have adapted their distribution structure to the presence of non-resident consumers (such as Cash & Carry stores). The 2005 Regional Accounting figures place the importance of consumption by non-residents at 30% of final domestic spending. Tourists consume food primarily in restaurants, with the amount of money they spend purchasing foods in retail stores being much less. The tourists, by and large, are from the north of Europe, meaning that their cultural affinities drive them to seek out foods that are "exotic" to them, while the local population is also confronted with "strange" foods on the shelves of supermarkets located in tourist areas.

## 3.2 Food supply

Having outlined the main characteristics of the food demand in the Canaries, we shall now summarise the trend in the supply, which is characterised by the gradual deagrarianisation of the economy of the Canaries throughout the 20th Century in valueadded, employment and foreign trade terms (see Figure 5). This trend has been accompanied in recent years by reduced production in absolute terms; from 2000 to 2010, the farmed area shrank by 7% and the output from agriculture, livestock and fishing dropped 14%. According to the Spanish Regional Accounts, primary sector employment in the Canaries decreased by 29% between 2000 and 2009.

Agriculture in the Canaries has a dual structure. On the one hand, the export sector is concentrated in a few products: bananas, tomatoes, ornamental plants and flowers, mainly in coastal areas and with a more capitalised and professional production system (Cáceres, 2000; Cid and Pérez, 2008; Nuez, 2005); on the other, the sector that targets its products for the internal market is characterised by the fragmentation into many products made in small, family-owned farms (Rodríguez and Gutiérrez, 1999; Consejo Económico y Social de Canarias, 2009). This duality is reflected in the fact that exported crops accounted for 34.1%

of the total farming production in 2010, while the livestock subsector, oriented mainly toward the internal market, represented 29.5%, primarily through the production of goat cheese and meat, pork meat and poultry eggs and meat.

| Indicator | 1985 | 1995 | 2000 | 2005 | 2010 |
|---|---|---|---|---|---|
| Agrarian GVA/total GVA (current prices) | 5.0 | 3.2 | 1.9 | 1.3 | 1.2 |
| Agrarian GVA/total GVA (2000 prices) | 3.5 | 2.9 | 1.9 | 1.3 | 1.3 |
| Agrarian jobs/total jobs | 15.7 | 8.2 | 3.3 | 2.3 | 2.0 (a) |
| Food exports (chap. 1 to 24)/goods and services exports | 65.7 (b) | 63.7 | 35.3 | 27.6 | 21.1 |

**Figure 5.** Standing of the agrarian sector within the economy of the Canaries, 1985- 2010 (percentages). (Sources: INE [Regional Accounts] and ISTAC [foreign trade]).

| Products | Value | % |
|---|---|---|
| AGRICULTURAL SUBSECTOR | 455,825 | 70.49 |
| Bananas | 130,292 | 20.15 |
| Tomatoes | 62,054 | 9.60 |
| Potatoes | 27,981 | 4.33 |
| Ornamental plants and flowers | 54,808 | 8.48 |
| Fruits | 59,562 | 9.21 |
| Wines | 13,428 | 2.08 |
| Other | 107,699 | 16.66 |
| LIVESTOCK SUBSECTOR | 190,811 | 29.51 |
| Meat | 63,014 | 9.74 |
| Milk | 63,859 | 9.88 |
| Eggs | 45,150 | 6.98 |
| Other | 18,788 | 2.91 |
| TOTAL | 646,636 | 100.00 |

**Figure 6.** Value of agricultural production (thousands of euros). Canaries, 2010. (Source: Council of Agriculture, Livestock, Fishing & Waters of the Government of the Canaries)

Insular differences in demographic, economic and agricultural terms are significant. The islands of Gran Canaria and Tenerife (Figure 7) are usually referred to as the 'central islands' not only due to their geographic location in the archipelago, but also to their contribution to the population and GDP, representing over 80% of both variables (Figure 8). Their contribution to agricultural production is also high, though considerably below their demographic and economic contribution, meaning that the agricultural activity in the so-called 'peripheral' islands outweighs their demographic contribution (Figure 9). The exception to this rule is Lanzarote, where both variables are similar. The largest differences, which reflect a trend toward food self-sufficiency, are seen in the smaller western islands (La Gomera, El Hierro, and, to a lesser extent, La Palma). This situation stems from two issues. The first is their reduced connectivity with the two central islands, through whose ports imported foods enter the Canaries. The second is the difficulty of connecting different parts within these islands due to their steep orography. Both aspects limit access to outside food supplies, and when it does arrive, it is considerably more expensive than in the rest of the archipelago. This forces local production to increase to supply the island. Much of the livestock in the relatively cattlerich areas of El Hierro and Fuerteventura is devoted to the production of cheese, with 5 to 10 litres of milk being required to produce one kilogram of cheese, a conversion that is not considered in the calculations used for Figures 8 and 9.

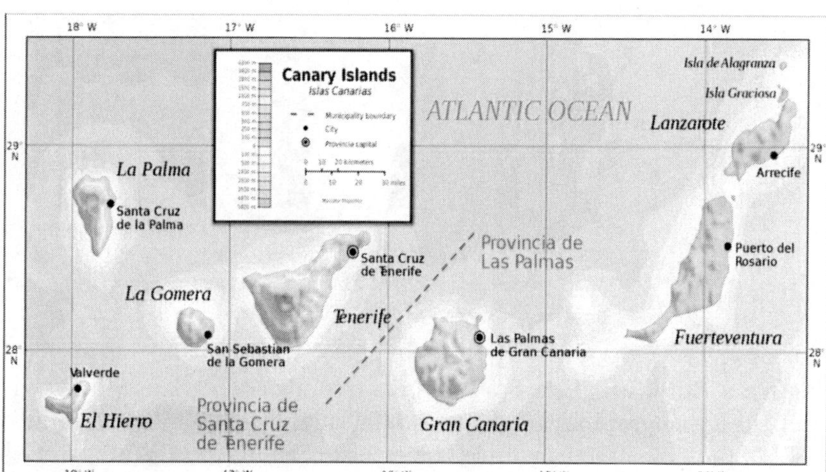


**Figure 7.** Map of the Canary Islands. Source: www.zonu.com

| Island | Demo-graphic weight 2008-10 (%) | % average GVApc 2009 | Weight in GVA 2009 (%) | Per capita agricultural production 2008-2010 (kg) (a) | Weight of agricultural production 2008-2010 (%) (a) | Per capita livestock production 2008-2010 (kg) (b) | Weight of livestock production 2008-2010 (%) (b) |
|---|---|---|---|---|---|---|---|
| Lanzarote | 6.7 | 109.6 | 7.2 | 181.1 | 7.4 | 69.7 | 5.4 |
| Fuerteventura | 4.9 | 116.8 | 5.3 | 45.1 | 1.3 | 345.0 | 19.5 |
| Gran Canaria | 39.9 | 98.6 | 39.9 | 193.2 | 39.6 | 84.5 | 39.0 |
| Tenerife | 42.8 | 99.6 | 42.4 | 124.9 | 31.7 | 51.7 | 25.6 |
| La Gomera | 1.1 | 100.7 | 1.1 | 405.5 | 2.7 | 178.4 | 2.2 |
| La Palma | 4.1 | 84.5 | 3.5 | 246.9 | 6.2 | 126.5 | 6.0 |
| El Hierro | 0.5 | 95.5 | 0.5 | 492.0 | 1.5 | 361.0 | 2.2 |

(a) Not including crops for export, industry and animal feed
(b) Not including other uses, mainly fertiliser

**Figure 8.** Basic insular indicators: demographics, economy, agricultural production. (Source: ISTAC, Council of Agriculture, Livestock, Fishing and Waters of the Government of the Canaries.)

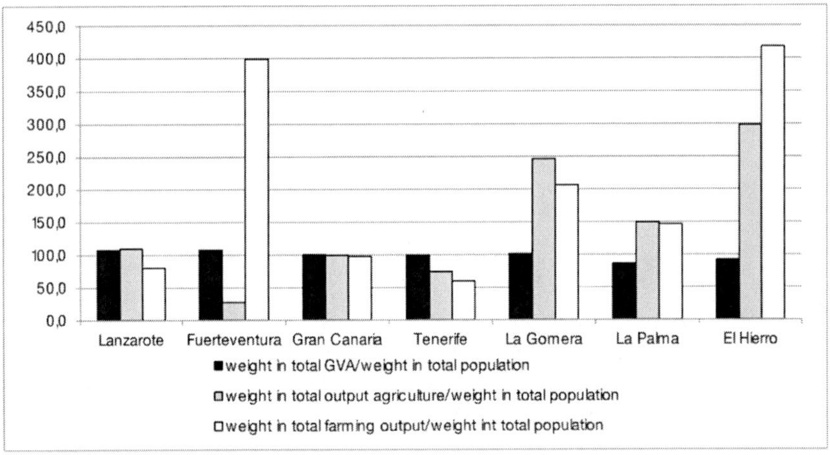


**Figure 9.** Ratios (in %) between the economic and agricultural weights and the demographic weight (triennial average 2008-2010). (Source: ISTAC, Council of Agriculture, Livestock, Fishing and Waters of the Government of the Canaries.)

The supply of food also includes the food industry. Its share of the scarce industrial activity in the Canaries is high. According to the Regional Accounts, in 2010 industry accounted for only 7.4% of the GDP of the Canaries, but 20.3% of its added value and 26.7% of jobs corresponding to Food, Drink and Tobacco. The relative importance of the food industry is explained by the effects of insularity. The low unit cost of food in comparison to other manufactured goods conditions the advantage of those industrial activities that eliminate low-value components (eg dehydration) in order to save on import shipping costs and reconstitute the food allocated for the local market. Examples of this are soft drinks and dairy products. In contrast, that portion of the food industry devoted to transforming local raw materials is underdeveloped and devoted mainly to products like wines and cheeses.

Since islands are our concern, a brief reference to fishing activity is in order. Ascertaining the apparent available supply poses certain complications involving statistical sources for non-manufactured goods, but not so for preserved or processed food. Since the Canaries are used to offload the fish caught in the nearby Saharan banks by the fleets of various nations for subsequent redirection as both frozen and fresh product, it is difficult to quantify how much of the offloaded fish remains in the insular market. In fact, the foreign trade statistics show a net outflow of fish that exceeds the local production figures of the Fisheries Department of the Government of the Canaries. Judging by the data of the Food Consumption Panel of the Spanish Ministry of Agriculture, Food and the Environment, the average amount of fish products consumed in households in the Canaries in the 2008-2010 period was 20.96 kilos per person per year, though exactly how much of that was produced locally could not be accurately determined.

## 3.3 Food self-sufficiency

Trying to measure the level of food self-sufficiency in the Canaries based on the apparent consumption (production + imports – exports) is no simple matter, but some estimates can be made that will yield at least one figure with which to evaluate the situation. In order to calculate the food and agricultural balance we used statistics from the Customs Office, available on the website of the Canaries Statistics Institute. This yielded the import and export figures, with the local production figures being provided by the Canaries Agricultural Statistics published annually by the regional government's Department of Agriculture, though neither source is entirely reliable (Nuez and Redondo, 2008).

To carry out the calculations, the foreign balance for fruits was calculated using tariff entries HSN0801 to HSN0810, while that for vegetables relied on entries HSN0701 to HSN0714, not including bananas, tomatoes and cucumbers, products that are earmarked primarily for export. Wine used entry HSN2204 and grains entries HSN10 and HSN11, with flour being converted to its grain equivalent. As for animal products, tariff entries HSN0201 to HSN0210 were used for meat, HSN0401 to HSN0406 for dairy, with cheese, butter and powdered milk being converted to their liquid milk equivalents. For eggs we added tariff entries HSN0407 and HSN0408, converted into fresh egg equivalents, while entry HSN0409 was used for honey. These calculations, then, excluded fish and fish products, beef and processed meats, food products and preserves made with vegetables, and breads and baked goods.

| | A. Local production (MT) | B. Foreign balance (MT) | C. Available supply (MT) | % A/C |
|---|---|---|---|---|
| Average 1990-1992 | | | | |
| Vegetables (a) | 223,455 | 64,223 | 287,678 | 77.68 |
| Fruits (b) | 89,961 | 112,126 | 202,087 | 44.52 |
| Cereals and flour (c) | 5,048 | 390,372 | 395,420 | 1.28 |
| Wine (d) | 17,800 | 47,575 | 65,376 | 27.23 |
| Average 2008-2010 | | | | |
| Vegetables (a) | 236,837 | 149,152 | 385,989 | 61.36 |
| Fruits (b) | 98,933 | 139,427 | 238,361 | 41.51 |
| Cereals and flour (c) | 2,190 | 384,993 | 387,183 | 0.57 |
| Wine (d) | 9,824 | 46,442 | 56,266 | 17.46 |

(a) Not including tomatoes and cucumbers for export, (b) not including bananas for export, (c) flour converted into its grain equivalent, (d) local grape production converted into wine

**Figure 10.** Apparent available supply of vegetable products (Tonnes). (Source: Council of Agriculture, Livestock, Fishing and Food of the Government of the Canaries. Customs Office. Compiled by authors.)

In the early nineties, there were only two sectors that, properly speaking, could be considered to exhibit a high level of self-sufficiency in the Canaries: egg and vegetable production. In both cases, the local supply accounted for more than 75% of the apparent consumption, with the egg supply meeting practically all of the demand. The situation involving other food products, however, was far different, with imported foods dominating the market. This situation continues to the present day, though the preponderance of local vegetable products is not as evident.

| | A. Local production (MT) | B. Foreign balance (MT) | C. Available supply (MT) | % A/C |
|---|---|---|---|---|
| Average 1990-1992 | | | | |
| Meat | 23.897 | 87.866 | 111.763 | 21,38 |
| Dairy (a) | 129.556 | 451.041 | 580.597 | 22,31 |
| Eggs (b) | 39.528 | 1.976 | 41.504 | 95,24 |
| Honey | 195 | 528 | 723 | 27,03 |
| Average 2008-2010 | | | | |
| Meat | 26.288 | 110.924 | 137.192 | 19,15 |
| Dairy (a) | 128.481 | 656.323 | 784.805 | 16,37 |
| Eggs (b) | 27.381 | 10.476 | 37.857 | 72,33 |
| Honey | 332 | 731 | 1.063 | 31,22 |

(a) Dairy converted into its liquid-milk equivalent, (b) egg products converted into their fresh-egg equivalent.

**Figure 11.** Apparent available supply of animal products (metric tonnes). (Source: Council of Agriculture, Livestock, Fishing and Food of the Government of the Canaries. Customs Office. Compiled by authors.)

In the case of eggs, the apparent consumption in the Canaries has dropped by 9%, in keeping with the general trend being seen in the entire Western world and attributable to changing consumption patterns. Local production has also decreased, by 31%, though it still accounts for 72% of the supply. As for vegetables, apparent consumption has practically matched the increase in population and is basically covered through imports. A second group is comprised by fruits and natural honey, where local production currently accounts for over 30% of the apparent available supply. The consumption of fresh fruits seems to have yielded some ground to other types of products, since the local supply has grown less than half as much as the population, with imported products once more gaining ground. In the case of honey, however, the amount of local product has increased by 70%, though it may be that the initial figure was underreported. The third block consists of meat, dairy and wine, the local production of which has less than a 20% market share and decreasing. In the case of wine, changes in consumption patterns have resulted in a reduction in the available supply of nearly 14%, in line with trends in the rest of the

Western world. This reduction has affected mainly the wines produced in the archipelago. As for meat, the available supply is also growing at a rate below that of the population, accounting for little over half. The remaining amount is covered with imports, since the increase seen in local production appears to stem from a statistical defect. As for dairy, despite initiatives carried out by public administrations starting in the 1960s to provide the archipelago with a livestock of dairy cattle to complement the existing livestock of goats, the results have been insufficient to meet the growing demand for dairy products. The increase in the available supply, therefore, which has practically matched the increase in the population, has been the exclusive result of foreign production. Finally, the presence in the market of locally grown cereals is purely anecdotal. Following the elimination of import tariffs in 1900, the cultivated land area in the Canaries started to decrease gradually to the point where it has practically disappeared due to the massive import of grains and flour from abroad at international market prices.

# 4. DETERMINANTS OF DECREASING FOOD SUFFICIENCY

The decreasing level of food self-sufficiency in the Canaries can be explained by a combination of factors that represent the interaction between supply and demand in a market with a particular institutional configuration.

The demand for food has risen as a result of both the population's increased size and purchasing power. The demographic evolution has been expansive, with a decisive migratory influx from abroad, particularly from European and Latin American countries. Part of the demand is also provided by tourism. The growing amount of land occupied by residential and tourism activities has contributed to the increased cost and scarcity of this production factor, making the prices of arable land in the Canaries the most expensive in all of Spain's regions (Ministry of Agriculture, Food and the Environment, 2011).

The increased purchasing power also contributes to this growing demand for food, particularly for those products with high demand-income elasticity, such as those with high protein content, fresh perishables and products processed by the food industry. In the case of foods consumed by tourists, in addition to the cultural particularities of their culinary customs, a high percentage of their food expenses is associated with the hotel industry and not with purchases at retail establishments.

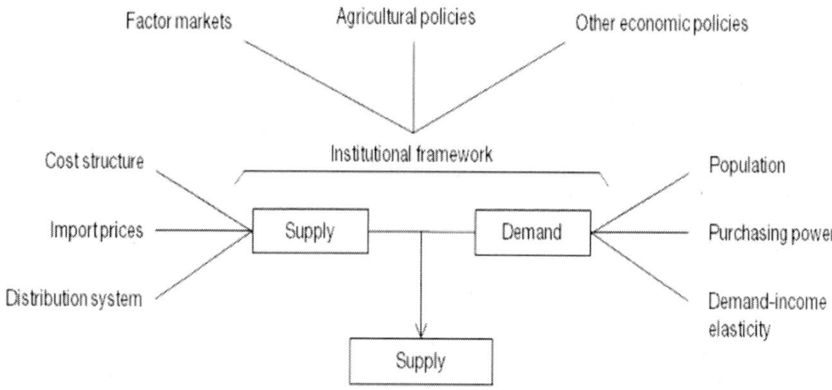


**Figure 12.** Determinants of food self-sufficiency

On the food supply side, the main aspect to consider is the competitiveness of the price of local products in comparison with imported foods. The competitive weakness of the local supply is manifested in more unfavourable cost structures resulting from a lack of resources (soil, water), the abrupt geography and the fragmentation of plots, as well as by inter-industry salary competition. The imported supply, in turn, must offset the cost of transporting these products to the Canaries, though the fraction of this cost in the final price is far outweighed by the cost disadvantages of local production. Particularly for products with a low unit value, the industrial transformation at the point of origin (eg dehydration) and destination (eg rehydration) contributes to reducing the impact of the costs of insularity.

Also, the high and growing concentration of retail stores run by multinational distribution chains poses the challenge of inserting local products into the shelves at these points of sale, which are supplied from large purchasing centres. The fragmentation of local production and its small horizontal integration in cooperatives devoted to the internal market in the Canaries hamper the commercial adaptation of local products to changes in the distribution network. Bearing in mind their limited price competitiveness, local products can only maintain or gain market share through non-price competitiveness, that is, by having the consumer accept their higher prices through product differentiation and market positioning. In the case of wines and cheeses, the price levels of quality local products are substantially higher than those of most imported products, a reflection of the success of this product differentiation in the eyes of the consumer. In the case of eggs, it is the added value of freshness that allows local production to maintain its place in the market in the face of cheaper imports. There are also untapped business opportunities to address the tourist demand for 'authentic' products, with the demand for agricultural tourism products being particularly unmet by a supply that scarcely conforms to the demands of tourism operators.

As concerns the institutional framework, the food and agriculture market is highly regulated and far from being 'free', given the importance society places on food safety and on protecting agricultural production. In the case of the Canaries, the islands are part of the European Union's Common Agricultural Policy (CAP) and its main export crops, bananas and tomatoes, are produced for the European market and protected under various regulatory mechanisms (Dupuis, 2003). If not for this privileged access to the European market, the presence of these crops in the Canaries would be reduced to a bare minimum.

The act of belonging to Spain and the European Union shapes the competitive profile of the food and agricultural policies of the Canaries. This institutional framework can be summarised by the unique arrangement of integration in the European Union. The POSEI programme (acronym for Programme of Options Specific to the Remote and Insular Nature of the Canary Islands), established by Council Decision 91/314/EEC, notes the need to apply a set of measures specific to the farming, fishing and commercial activities in the Canaries while safeguarding certain customs and tax aspects that have characterised the economic history of the islands.

The aim of the POSEI programme is to increase the share of the primary local sector in meeting food needs through subsidies to production and marketing. Thus, an aid programme is available per kilo of fruits and vegetables produced by those farmers who sell their products in the local market. The cultivation of potatoes is subsidised both by hectare and by yield marketed. Aid per hectare of vineyard is also available for the production of wines with protected appellations, as are subsidies for the processing and bottling of these wines. As for livestock products, the import of breeding stock is subsidised, along with the birth and slaughter of rams, the raising of goats and sheep, along with the commercialisation of milk, meat and eggs obtained from the islands' farms. To these measures we must add the Rural Development Programme subsidies aimed at improving competitiveness in the agriculture and livestock sectors, the main objectives of which are to modernise both farming and livestock operations as well as to provide increased value to farm products.

| Direct aid in 2010 | Thousands of euros |
|---|---|
| Total payments from the European Agricultural Guarantee Fund (EAGF) (a) = total aid from POSEI | 268,400 |
| - Specific supply arrangement | 72,700 |
| - Vegetable production subsidies | 34,600 |
| - Banana subsidies | 141,100 |
| - Animal production subsidies | 20,000 |
| Rural Development Programmes (RDP) (b) | 19,300 |

(a) Payments with no modulation applied. The annual financial outlay is predefined in Article 23 of Regulation (EC) No. 247/2006.
(b) For programmes for the rural development of the Canaries and Spain: European Agricultural Fund for Rural Development (EAFRD) + national expenditure + regional expenditure.

**Figure 13**. European Union aid to agriculture in the Canaries. Source: Spanish Agricultural Guarantee Fund (SAGF). (Taken from Massot, 2011: 27.)

In spite of this objective, and indicative of the commercial sector's increased political power, production intended for export has managed to capture most of the POSEI aid. Of the programme's 268 million euros, 141 are earmarked for banana production, the main market for which is on the Spanish mainland. Likewise, of the 34 million in aid for vegetable production, 7 are for the commercialisation of fruits, vegetables and plants outside the Canaries, meaning that only 18% of EU aid to the farming industry in the Canaries is used to increase production for the local market.

Since an increase in the share of local production is practically unviable for some products, a Specific Supply Arrangement (SSA) was created to reward the importation of certain products (raw materials for the food and agriculture industry) for the purpose of balancing the cost of food in the Canaries with that typical on the European mainland. The effects of these measures have been contradictory. On the one hand, there has been an increase in the local production of fruits and vegetables. On the other, while the importation of cheaper cereals for the manufacture of animal feed so as to improve the livestock has benefited the income of local producers, the SSA subsidies have also made it possible for meat, milk and dairy products to enter the insular market in amounts and at prices that limit their marketing possibilities, and with it their growth (Redondo, 2009: 362).

Within the institutional framework affecting agricultural activity in the Canaries, a reference to the markets for the production factors (land and water in particular) is in order. In the Canaries, water is a scarce and privately-owned resource, meaning that historically its use has been diverted to those production sectors that were best able to pay for it (Aguilera, 2002; Nuez and Carnero, 2003). In the past this made it difficult for the agricultural production intended for the local market to access the resource since crops for export absorbed most of the available water. Today it is competition from urban and tourist areas that hampers its use for agricultural activity. In the real estate market, the most relevant institutional elements to agricultural production are the lenient rural zoning regulations in the face of pressure from urban development, the presence of wide-ranging natural protected areas with limited access to agricultural production, and the possibility of purchasing arable land for non-production purposes (speculation for future resale in a change-of-use context).

# 5. PERSPECTIVES FOR IMPROVING LOCAL PRODUCTION

The low level of food self-sufficiency in the Canaries is generating increasing social concerns (Bermúdez, 2007; González and Santana, 2007; Redondo, 2010). The stagnation, if not reduction, in the volume of food produced in the archipelago is the result of a set of factors in which political action/inaction plays a very important role. Despite the existence of numerous reports written by or for various public organisations, little has been done in their wake to halt or reverse the situation.

In 1996 the Parliament of the Canaries unanimously approved the final report of a commission to study the Specific Supply Arrangement, and which clearly stated how the level of self-sufficiency was regarded as excessively low. Godenau et al. (1997), in their report for the Government of the Canaries' General Directorate for European Affairs in the effect of the SSA on the archipelago's food and agricultural system, posed the problems in the application not only of that instrument, but of the entirety of the measures included in the POSEI programme. This same point was made two years later in a study for the Government of the Canaries' General Directorate for Economic Promotion (Martínez, 1999). The 2002 annual report of the Economic and Social Council of the Canaries (Consejo Económico y Social de Canarias, 2003) also noted the market distortions resulting from the application of that assistance. This same organisation's 2008 report featured a section specifically addressing the agricultural and livestock situation in the Canaries that reiterated these distortions. More recently, a report prepared by the consultancy Oréade-Brèche (2009) for the European Commission noted the need to lower the level of aid and the volume of the SSA forecast plan for products that compete with local production in the Canaries.

Based on the premise that there are foods whose demand in the Canaries could not be met solely through local production in the archipelago, or if so only at costs that would make meeting such demand unviable (Machín and López-Manzanares, 2011), increasing the supply of a large number of locally produced items should be considered so as to keep the level of dependency on outside goods in check. There are several possible courses of action:

- First, in the area of agricultural policy, the practice of resorting to community policies (POSEI and RDP) could be shifted to place more emphasis on production intended for the internal market, thus giving less importance on the commercial aspect that supports imports through the SSA of foods that compete with local production intended for the internal market and specifically aids export crops.

- Second, industrial policies could be used to selectively promote the industrial transformation of local agricultural products. Such policies have been relatively successful with wines and cheeses, and the potential remains to diversify these local industries.

- Third, given the weakness in marketing local foods, their presence in distribution channels must be encouraged, paying particular attention to short channels and the 'experience economy' (Pine and Gilmore, 1998; Godenau, 2007). The advantage of the proximity of local products should be actively communicated as part of food and agricultural marketing (European Network for Rural Development, 2012). More favourable carbon footprints than for imported products and more direct contact with the producers and the manufacturing processes are just some examples of how the 'experience economy' can improve the non-price competitiveness of local products. Specific measures can consolidate the agrotouristic cluster.

- Lastly, in the legal-regulatory arena, support measures should reduce urban sprawl, reinforcing the effective implementation of a regulatory framework that is sufficiently powerful to protect agricultural land. Also of interest are modifications to the tax regime governing family farms and supporting the possibility of allowing families residing in urban areas to develop peri-urban agriculture. In general terms, access to the means of production (soil, water, financing) poses a significant obstacle to the development of new entrepreneurial initiatives in the industry.

The Canaries' history of agricultural exports and trade continues to weigh heavily on today's agricultural policies. The attention paid in recent decades to foods intended for the local market has not been enough to offset the disadvantages associated with production costs. Nevertheless, the key opportunities for the future of agricultural activity lie in the archipelago's internal market and not in exports. Building the structure for an inter-insular market, taking advantage of the proximity of local production and the high quality of some of the products made are arguments that cast the future of food self-sufficiency in a favourable light.

## END NOTES

[1]Moreover, it is typical to use States when delimiting the units being compared. Since islands have a high propensity to be Small Insular

States, the average size of mainland States is bigger (Taglioni, 2011:58).
[2] Potatoes boiled with their skin in very salty water.

[3] Apparent consumption implicitly includes the HORECA sector and personal consumption, but there are other factors that limit a comparison between the two ways of estimating market volume. Apparent consumption includes consumption by nonresidents (tourists) but overestimates the local production of certain products due to record management deficiencies involving cultivated land area.

# BIBLIOGRAPHY

1. Aguilera Klink, F (2002) Los mercados de agua en Tenerife ('The water markets in Tenerife'), Bilbao: Bakeaz

2. Amann, S, Haunhorst, C, Ludwig, U et al (2012) 'Diarrea china en la UE' ('Chinese Diarrhea in the European Union'), El País, 28 October, online at: www.elpais.es - accessed July 2013

3. Baldacchino, G (2004) 'Autonomous but not sovereign? A review of island subNationalism', Canadian Review of Studies in Nationalism v31: 77-91

4. Barnett, J (2011) 'Dangerous climate change in the Pacific Islands: food production and food security', Regional Environmental Change v11 (Suppl 1): S229–S237

5. Bermúdez Hernández, J.J (2007) ¿Por qué duerme la población y el sector agrícola ante la extinción de la actividad agropecuaria canaria? ('Why does the population and the agricultural sector sleep while facing the extinction of the Canarian farming industry?') – online at: decrecimientoencanarias.blogspot.com.es/2007_06_01_archive.html - accessed July 2013

6. Burriel de Orueta, E (1982) Población y agricultura en una sociedad dependiente '(Population and agriculture in a dependent society'), Barcelona: Oikos Tau

7. Cáceres, J.J (2000) La exportación de tomate en Canarias: Elementos para una estrategia competitive ('The Canary Islands' tomato exports: Elements for a competitive strategy'), Santa Cruz de Tenerife: Ediciones Canarias

8. Chapman; KR (2009), Rapid Appraisal of the Bioenergy-Food Security Nexus in Pacific Island Countries, FAO Sub-Regional Office for the Pacific Islands, Final Report, online at: www.sprep. org/ att/irc/ecopies/pacific_region/498.pdf - accessed July 2013

9. Cid, M.C and Pérez, M.L (2008) 'El sector ornamental de Canarias' ('Canary ornamental sector'), Horticultura, n204: 60-69

10. Consejo Económico y Social de Canarias (2003) Informe anual del CES, 2001-2002 ('Annual report of the Canary Islands Socioeconomic Council, 2001-2002'), Las Palmas de Gran Canaria, online at: www.cescanarias.org - accessed July 2013

11. (2009) Informe Anual 2008 '(Annual report 2008'), Las Palmas de Gran Canaria, online at: www.cescanarias.org – accessed July 2013

12. Contreras Hernández, J and Gracia Arnáiz, M (2005) Alimentación y cultura: Perspectivas antropológicas ('Food and culture: Anthropological perspectives'), Barcelona: Ariel

13. Dupuis, I (2003) Los efectos de las políticas comunitarias sobre las agriculturas de las regiones ultraperiféricas: el caso de las Islas Canarias ('The effects of EU policies on agriculture in the outermost regions: the case of the Canary Islands'), La Laguna: Ediciones Baile del Sol

14. European Network for Rural Development (2012), Local Food and Short Supply Chains, EU Rural Review, n12, Summer 2012

15. FAO (2002) The state of food insecurity in the world 2001, Rome: FAO, online at: http://www.fao.org/docrep/005/ y4671e/ y4671e06.html - accessed July 2013

16. (2008) Climate Change and Food Security in Pacific Island Countries, Rome: FAO

17. Godenau, D (2007) 'El vino y la economía de las experiencias' ('Wine and the experience economy'), Vinaletras, Cuaderno bianual de cultura y vino TacoronteAcentejo v1: 46-57

18. Godenau, D et al (1997) Informe sobre la incidencia del REA en el sistema agroalimentario ('Report on the incidence of REA in the agri-food system'), Santa Cruz de Tenerife: Dirección General de Asuntos Europeos del Gobierno de Canarias

19. González Cortés, M.A and Santana Rodríguez, J.M (2007) 'Una Reserva Estratégica Alimentaria para Canarias' ('A Strategic Food Reserve for the Canary Islands'), Información Veterinaria, Septiembre: 20-24

20. Gough, K.V, Bayliss-Smith, T, Connell, J and Mertz, O (2010) 'Small island sustainability in the Pacific: Introduction to the special issue', Singapore Journal of Tropical Geography v31: 1–9

21. Gutiérrez Hernández, P and Godenau, D (2010) 'Oferta y demanda de servicios comerciales minoristas en Canarias' ('Supply and demand for retail commercial services in the Canaries'), in Montero Muradas, I and Gutiérrez Hernández, P (dirs), La actividad comercial en Canarias: Dimensiones estructurales y empresariales ('Commercial activity in the Canary Islands: Structural and business dimensions'), Santa Cruz de Tenerife: Fundación FYDE-CajaCanarias: 45-73

22. King, R (2009) 'Geography, Islands and Migration in an Era of Global Mobility', Island Studies Journal v4 n1: 53-84

23. Machín Barroso, N and López-Manzanares Fernández, F (2011) Autoabastecimiento Agrícola en Tenerife. Implicaciones territoriales '(Agricultural Self-sufficiency in Tenerife:

24. Territorial implications'), Santa Cruz de Tenerife: Servicio Técnico de Agroindustrias e Infraestructura Rural, Cabildo Insular de Tenerife

25. Macías Hernández, A.M (1981) 'El papel histórico de la agricultura de «subsistencia» en Canarias: un tema olvidado' ('The historical role of subsistence agriculture in the Canary Islands: a forgotten issue'), in Canarias ante el cambio ('The Canary Islands facing change'), Instituto de Desarrollo Regional, Universidad de La Laguna, Banco de Bilbao, Junta de Canarias: 101-112

26. (2001) 'La construcción de las sociedades insulares: el caso de las Islas Canarias' ('The construction of island societies: the case of the Canary Islands'), Estudios Canarios. Anuario del Instituto de Estudios Canarios [2000], XLV: 131-160

27. Martínez Muñoa, A (1999) El Régimen Específico de Abastecimiento de las Islas Canarias. Campañas 1992/93 a 1998/99 ('The specific supply arrangements of the Canary Islands: Campaigns from 1992/93 to 1998/99'), Santa Cruz de Tenerife: Dirección General de Promoción Económica de la Consejería de Economía y Hacienda del Gobierno de Canarias

28. Massot, A (2011) La agricultura de las islas Canarias ('Agriculture in the Canary Islands'), Dirección General de Políticas Interiores de la Unión, Departamento Temático B: Políticas Estructurales y de Cohesión, Agricultura y Desarrollo Rural, documento solicitado por la Comisión de Agricultura y Desarrollo Rural del Parlamento Europeo

29. Mertz, O, Bech Bruun, T, Fog, et al (2010) 'Sustainable land use in Tikopia: Food production and consumption in an isolated agricultural system', Singapore Journal of Tropical Geography, v31: 10–26

30. Ministerio de Agricultura, Alimentación y Medio Ambiente de España (2011), Encuesta de precios de la tierra 2011 '(Survey of land prices 2011'), archived online at: www.magrama.gob.es/es/estadistica/temas/estadisticasagrarias/Encuesta_Precios_Tierra_2011_tcm7-220709.pdf

31. North, D (1990) Institutions, institutional change and economic performance, Cambridge: Cambridge University Press

32. Nuez Yánez, J.S (2005) El mercado mundial de plátanos y las empresas productoras en Canarias, 1870-2000 ('The world banana market and production companies in the Canary Islands, 1870-2000'), Santa Cruz de Tenerife: Instituto de Estudios Canarios, ASPROCAN y Caja Rural

33. (2008) '¿Qué vino hemos bebido? Caracterizando la oferta disponible de vino en una economía abierta: Canarias, 1886-2006' ('Which wine did we drink? Characterising the available wine supply in an open economy: Canary Islands, 1886-2006'), Territorios del Vino n3: 38-64

34. Nuez Yánez, J.S and Carnero Lorenzo, F (2003) 'El mercado del agua en Canarias: una perspectiva histórica' ('The water market in the Canary Islands: A historical perspective'), Revista de Historia Económica vXXI n2: 373-398

35. Nuez Yánez, J.S and Redondo Zaera, M (2008) 'La Balanza Agroalimentaria de Canarias' ('The Canary Islands' agri-food balance'), Hacienda Canaria n24: 49-80

36. Oréade-Brèche (2009) Évaluation des mesures mises en œuvre en faveur des régions ultrapériphériques (POSEI) et des petites îles de la mer Egée dans le cadre de la politique agricole commune '(Evaluation of measures implemented in the outermost regions (POSEI) and the small islands of the Aegean Sea under the Common Agricultural Policy'), online at: ec.europa.eu/agriculture/eval/reports/posei/index_fr.htm. - accessed July 2013

37. Pine, B.J and Gilmore, J.H (1998) 'Welcome to the experience economy', Harvard Business Review, July-August: 97-105

38. Redondo Zaera, M (2009) 'Dudas y certezas sobre el sector agrario orientado al mercado interior' ('Doubts and certainties about the domestic market-oriented agricultural sector'), Rincones del Atlántico n6/7:352-367

39. Redondo Zaera, M (2010) 'Autoabastecimiento y soberanía alimentaria en Canarias' ('Self-sufficiency and food sovereignty in the Canary Islands'), Mundo Rural de Tenerife n9: 8-11

40. Rodríguez Martín, A and Gutiérrez Hernández, P (1999) 'La agricultura en las Islas Canarias' ('Agriculture in the Canary Islands'), in Molina Ibáñez, M et al (coordinators) El sector agrario. Análisis desde las Comunidades Autónomas ('The agricultural sector: Analysis from the Autonomous Communities'), Madrid: Ministerio de Agricultura, Pesca y Alimentación, Ediciones Mundi-Prensa: 119-150

41. Sans, J.A (1977) La crisis de la agricultura en Canarias ('The agricultural crisis in the Canary Islands'), Las Palmas de Gran Canaria: Excma. Mancomunidad de Cabildos de Las Palmas

42. (1981) 'Sobre la funcionalidad de la agricultura canaria en el desarrollo económico' ('About the functionality of the Canarian agriculture in economic development') in Canarias ante el cambio ('Canary Islands facing change'), Instituto de Desarrollo Regional, Universidad de La Laguna, Banco de Bilbao, Junta de Canarias

43. Taglioni, F (2011) 'Insularity, political status and small insular spaces', Shima: The International Journal of Research into Island Cultures v5 n2: 45-67

44. United Nations (1998) Development of a vulnerability index for small island developing States, Report of the Secretary-General, United Nations General Assembly, Economic and Social Council Environment and sustainable development: implementation of the outcome of the Global Conference on the Sustainable Development of Small Island Developing States, online at: islands.unep.ch/d98-vul.htm – accessed July 2013

45. Warrington, E and Milne, D (2007) 'Island Governance' in Baldacchino, G (ed) A World of Islands: An Island Studies Reader, Charlottetown, Canada and Luqa, Malta, University of Prince Edward Island, Institute of Island Studies and Agenda Academic: 379-427

5

# TOURIST CONSUMER BEHAVIOUR INSIGHTS IN RELATION TO ALL INCLUSIVE HOTEL RESORTS: THE CASE OF ANTALYA, TURKEY

**Tamer Ayik[1], Theodore Benetatos[2] & Ioannis Evagelou M.Sc.[3],**

[1]University Centre Switzerland

[2] University Centre Switzerland, Postgraduate Programme Leader

[3] Lecturer IMI University Centre Switzerland

## ABSTRACT

This research investigates tourist customer satisfaction levels in all-inclusive resort hotels in Turkey. The paper investigates changes in patterns regarding all-inclusive hotel packages as well as changing consumer behaviour of tourists preferring such establishments. The research was conducted in the region of Antalya and data is analysed by using SPSS with descriptive statistical analysis, and correlation analysis. The results suggest, among other, that new tourists that purchase all inclusive package are now more diverse in profile while the products themselves are significantly better than what all-inclusives were associated with. Relevant criticism is being addressed and suggestions for further research are being made. Finally, the study emphasized the effect of all inclusive system on customer satisfaction in 5 star hotels in Antalya.

## KEYWORDS

Consumer Behaviour, All-inclusive Resorts, Tourism Planning, Tourism


## 1 INTRODUCTION

The on-going financial crisis, initially after 2008 in the US and lately in the EU zone has created new playing fields and norms in the hospitality industry. Although it is relatively early to draw conclusions, it seems that hospitality firms exposed to debt were severely damaged and traditional tourism markets, such as destinations in the Mediterranean, found themselves in an interesting but also dramatic equilibrium between trying to be competitive and deliver quality products and services. Large firms tried to cut their losses while SMEs struggled simply to continue to exist in an environment where the cost of capital remained expensive and average hotel prices remained significantly lower than the 2008 mark (Hotel Price Index Data, 2012). However, the tourism and hospitality industry has shown great reflexes in the following years allowing global Travel & Tourism direct contribution to GDP to increase by a healthy 3.2% in 2012 (WTTC, 2013) but this does not reflect equally the majority markets. Political turmoil in Africa has enabled traditional markets like Greece and Spain or the nearby Dubai to maintain or even increase their competitive position mainly due to displaced demand (Ernst & Young, 2012). However, it seems that Turkey will outperform the Mediterranean resort market capitalizing on long term effects of the devaluation of the lira which has made the Turkish tourism product very attractive in comparison to Eurozone markets (Ernst and Young, 2012). According to the same source, it seems that this projected growth is mostly revolving around the *Sun and Sea* all-inclusive offerings that Turkey is well known for and which is a growing segment on a global scale, despite the fact that it is not a new product. It is however a new boost of all-inclusive facilities supply which provided the segment with a second youth. These interesting developments are receiving old criticism which still applies since authors have argued in the past about how all inclusive packages bode ill for many small business people - restaurant owners, craft vendors, taxi drivers - establishments owned by small business people (Henry, 1989). Critics of all-inclusive resorts accuse the firms for alienating the tourists from the local context by importing provisions and discouraging them in many ways tourists from ever leaving the properties (Issa and Jayawardena, 2003). This creates an adverse effect on the expected outcomes of tourism development. Furthermore it has been argued that early all-inclusive practices were not up to the desired standards with the adoption of unified pricing system leading to a decline in employee satisfaction as well as customer perception of overall service quality, and finally, in customer satisfaction (Kozac, 2006). Further evidence has shown that hotels around the globe are increasingly adopting all-inclusive policies and it has been documented that this trend has reduced the revenue per stay as well as the customer satisfaction levels (Aguilo and Rosselio, 2012).

However, evidence shows that times have changed and practices vary between properties. Increasing volume of tourists preferring all-inclusive packages might show a shift either in the quality of the delivered services or the tolerance of the tourists who might be willing to compromise the quality for a good price. Based on these prevailing conditions, the main objective of this research is to investigate consumer perceptions on services provided by all-inclusive resorts. Turkey as a premium and growing all-inclusive provider has been chosen for the research with a specific focus on the region of Antalya.

## 2 LITERATURE SUPPORT

Over recent years, several studies have focused on tourist satisfaction in particular tourism destinations and from there information was drawn on the elements that create attractive tourism products and services. This study concentrates on customer satisfaction with the service performance of all inclusive resort seeking to identify key variables which are related to customer satisfaction and holiday selection patterns.

The basic definition of all-inclusive holidays is where all travel, accommodation, food and beverage, entertainment, trips or sports coaching costs are paid for when booking for one established price. Issa and Jayawardena (2003), established that the all-inclusive holiday concept was first introduced in holiday camps in Britain in the 1930's and was a different holiday concept where a guest had to carry no form of currency while on a holiday. It was a package designed to eliminate extra charges that can sour the sweetest of vacations.

Issa and Jayawardena (2003) further explain that all-inclusive clubs had a major impact on the concept of hospitality service. In the 1950's,

'Club Mediterranean', popularly known as 'club-med' introduced the concept on the Spanish island of Mallorca and later spread all over the globe. An explosion of the concept was later realised along coastal strips in the late 1990s. At this point, the all-inclusive concept was a fundamentally different type of holiday where a guest had to carry no form of currency, paper, coin or plastic to a hotel, resort or while on a cruise. Poon (1998) describes the all-inclusive concept as an important product innovation in the international tourism market place, as it packages various elements of the tourism product. The main idea of all-inclusive system is, minimize monetary transactions during the holiday, do not to make some decisions about money spending during the holiday and to eliminate the anxiety because of spending money, stress or problems for tourists (Poon 1998). The all-inclusive system has so far been dominated by European based providers such as Italy, Greece, Turkey and France. Additionally, the Caribbean, Central America, Dominican Republic and other tourist destinations such as Thailand have entered the market with accommodation facilities (Menekse, 2005). German and Italian tourism enterprises have played a key role in spreading all-inclusive system in Europe, Asia and Africa. For example, main tour operators, such as TUI, NUR, ITS, preferred to implement all inclusive system in tourist destinations, such as, Spain, Turkey, Greece, Italy, Austria, Switzerland, Kenya, Sri Lanka, the Canary Islands, Tunisia, Senegal, Thailand and Ibiza with resorts operated by themselves like Robinson Club, Club Calimera and Club Aldiana (Poon 1998; Voss 2003).

This development has led to increased research interest mostly around motivational factors driving tourist consumer behaviour. Literature for some years now has recorded various factors which have impacts on consumers' purchasing decisions, such as cultural, social, personal and psychological (Mucuk, 1998). Other studies provide generally accepted classification of factors influencing consumer behaviour collected in four main groups such as personal factors (age and life cycle, occupation, economic status and lifestyle), cultural factors (culture, subculture, social class), social factors (reference groups, family, roles and status) and psychological factors (motivation, perception, learning, beliefs). These factors cannot be controlled by marketers generally, but necessarily be taken into account and to be emphasized on these factors (Kotler, Bowen, and Makens, 2006).

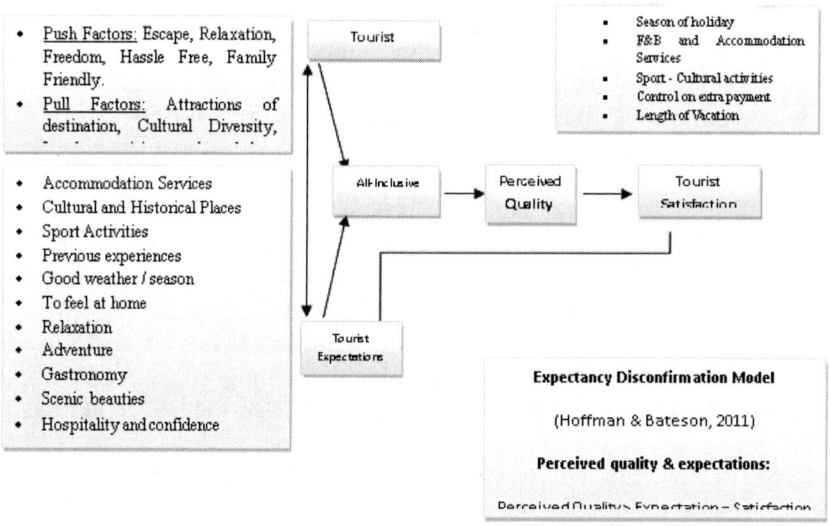


**Figure 1.** Tourist Motivational Factors and consumer decision making processes in relation to all-inclusive travel packages

All these factors affect tourist decision making process. These decisions can appear in a simple structure or sometimes, be connected with multiple decisions in more complex structures (Decrop 2006). These processes usually consist of the following steps of; identification of variables of the problem addressed by a purchase, evaluation of alternatives as solutions to the problem, and the related returns, and finally selecting the best alternative using the appropriate decision criteria to get the best result (Clemen and Reilly 2001). Consumer behaviour models emphasize that needs and motivation factors are closely associated with each other (Kozak 2002). Additionally motivation factors explain how tourists are pulled by various factors when deciding on their purchase and how tourists are pushed towards their holiday destination (Baloglu and Uysal 1996; Kozak 2002; Yoon and Uysal 2005).

# 3 RESEARCH AND METHODOLOGY

For the purpose of this study satisfaction levels were measured exclusively in 5-star all-inclusive hotel establishments in the highly touristic province of Antalya which has one of the most widely all-inclusive systems in Turkey. The main research questions to be answered where 'Why do the tourists prefer All- inclusive system?' determine the satisfaction level towards to the hotel services and features and last to identify the comparison with all-inclusive 5 star hotels with standalone 5 star hotels.

This research applied mostly quantitative research methods in consequence of the means available as well as the nature of this study. A survey of 18 questions was employed and 10 of the survey questions were close ended. The questionnaire contained 4 main sections. The first section was designed to determine the demographic characteristics of tourists. The second section measured satisfaction levels on food and beverage services, staff performance, and the general features of the hotel during the tourists stay. The third section investigated the importance of holiday destination features in decision making. The fourth section was designed to determine the importance of selection elements related all-inclusive resort hotels. Generally 5-point Likert-type scales were used and data obtained from the study were analyzed with SPSS (Statistical Program for Social Sciences). Demographics have been considered and the significant data have been continually evaluated. Reliability was also tested with Cronbach's Alpha Reliability test.

## 3.1 Sampling and Hypotheses

The study sample was chosen from 8 different 5 star hotels in the province of Antalya to represent the best of the population. According to data from the Ministry of Culture and Tourism, Turkey in 2011, 7.5 million tourists had visited in Antalya. Approximately 70% of tourists staying in Antalya prefer all- inclusive properties. To assume reliability a confidence level of 95% with a sample of 384 tourists would be ideal. However, we managed to gather 174 usable surveys which poses certain limitations but could be still considered an acceptable sample size for the needs of this research. The study sample was randomly selected.

According to the study aims certain hypothesis were created to identify the relationships between demographic characteristics of tourists and tourist's satisfaction with all inclusive resort hotel services. The hypotheses are:

H1.0: There is no relationship between education level and reason to choose All-inclusive holiday.

H1.1: There is relationship between education level and reason to choose All-inclusive holiday.

H2.0: There is no relationship between gender's food and beverage department satisfaction and overall satisfaction.

H2.1: There is relationship between gender's food and beverage department satisfaction and overall satisfaction.

H3.0: There is no relationship between the education levels of tourist with comparison the all-inclusive hotels quality to standalone 5 stars hotels quality.

H3.1: There is relationship between the education levels of tourist with comparison the all-inclusive hotels quality to standalone 5 stars hotels quality.

H4.0: There is no relationship between reasons to choose an all-inclusive resort hotels and involving any activity outside of hotel during the stay.

H4.1: There is relationship between reasons to choose an all-inclusive resort hotels and involving any activity outside of hotel during the stay.

# 4 ANALYSIS

The sample demographics were appropriate with fair age and gender distribution. The largest age range is 36-45 age group with 43,1% 26-35 age group is second group with 24,1%. Furthermore, 46-55 age range has 23,0%, 18-25 age group has 5,7% 56 and above age group has 4,0%. The Gender distribution was almost a 50-50 split which is also representative in nature.

The geographical origin distribution of the sample poses slight limitations to this study since the distribution was not ideal. The vast majority of the respondent were Germans representing 44,3% . The next big segment were the Russians with 19,0% and the Ukrainians with 9,2%. From there on 5,2% were from Austria, 4,0% from

Switzerland. On the positive side tourists from 18 different countries were surveyed in total.

The marital status was also even with 49,4% being married. Interesting enough 74.1% had no children and this was not expected in all-inclusive establishments but this could be justified from the age band of the interviewed tourists.

Another very interesting element refers to the education level which seems to be exceptionally high. 48,3% had BA university degree and amazingly 19,5% had doctoral degree. This is very interesting since it has been the impression in the past that well educated tourists prefer more independent travel products so this should be further investigated. 22,4% had high school degree and 9,8% had primary school degree.

The next set of questions aimed in identifying the level of satisfaction on Food and Beverage quality, staff performance, hotel facilities and crucial factors that affect purchase decisions. In total 24 different factors were investigated and the results were again somewhat surprising. It has been argued that in recent literature, that all-inclusive establishments suffer in terms of quality and (Aguilo and Rosselio, 2012) therefore brand loyalty but clearly here this is not the case. 51,1% were strongly satisfied with freshness of food while 52,9% were strongly satisfied of the food which was served in food and beverage outlets during their stay in 5 star all inclusive resort hotels. Menu descriptions and diversity of beverages have more satisfied tourists than strongly satisfied tourists. In total none of the participants were dissatisfied or strongly dissatisfied with any of the aspects of the food and beverage services.

Similarly tourists were satisfied with the performance of staff with none of the respondents providing a negative answer. Having said that some of the answers were at the neutral zone which could an indication of indecisiveness of what good performance of staff means or there was a struggle in compromising the quality level for a good deal. In the same fashion Hotel facilities were also very highly rated with total satisfaction levels that borderline the average of an 85% (strongly or just satisfied). Interesting enough the lowest scores went to children facilities rates.

Concluding this part of the research, important elements for holiday choice were investigated and the outcome was more or less the expected. *Price*of hotel is the most important feature for the survey participants with 88,5% specifying that it is very important element for choosing holiday product. Second most important factors are *Relaxation* and *Rest* with a percentage of 72, 4%. Amazingly 67, 2% did not consider the *Location* of hotel as an important element for deciding holiday which can be interpreted in various ways. It seems that the effect of choices provided by tour operators might render destination irrelevant but this is just an assumption and should be further investigated. What we can safely say though is that this proves that some of the on-going criticism presented earlier seems to apply. To further these arguments only 3, 4% seek for both amenities of destination and for Cultural and Historical places as important elements for deciding on holiday destination.

With regards to distribution channels used the analysis shows that 32,8% made their reservations through the travel agencies and 31,0% via tour operators. This means that 63,8% use the specific channels which illustrates a fairly high dependency on this means. Internet bookings were at 17,8% and this could also refer to the same as above channels of all-inclusive resorts. Finally, direct hotel bookings were at 18,4% which illustrates the early quoted dependency on certain channels and might also explain the high concentration of certain markets that seems to be key clients (Germany).

Finally a very important section of the research revolved around the identification of elements that lead tourist to all-inclusive resorts. The key motivational factor here is *Low Extra Expenses* that all-inclusive resort hotels offer to the tourist with a percentage of 84,5%. 81% of the respondents specified that no obligation for tracing the extra expenses is another very important reason to stay in all inclusive resort hotels. Hence we can conclude that tourists more focused on initial price and overall cost at the end of their vacation. The third most important motive is not to have to engage with holiday plans standing at 59,8%. Interesting enough 36,8% specified that it's of no significance to them if the establishment is family friendly but this can be possibly explained by the demographics and could vary in different destinations. Also to reject some of the criticism discussed, more than 90% were satisfied with the whole experience 46,0% of which were very satisfied and none was dissatisfied.

This first image of the analysis was somewhat expected but interesting findings were made and conclusions were drawn. However, the final part of the analysis refers to the investigation and testing of the stated hypotheses.

*H1.1: There is relationship between education level and reason to choose All-inclusive holiday.*

As mentioned previously, there were 7 elements of importance to choose all-inclusive resort hotels. These were, low extra expenses, all inclusive resort hotel services, being family friendly, sun, sea and sand features of the destination, no obligation for tracing the extra expenses, being holiday plan is ready and sport activity diversity. With cross tabulation we found that there is a relationship between education and perceived low extra costs which is probably connected also to the income level of the more educated respondents. On the Chi-square test, P-value was found 0.00. Since the P-value is between $0.00<0.05$ which is true for all the elements so H1 hypothesis is accepted.

*H2.1: There is relationship between gender's food and beverage department satisfaction and overall satisfaction.*

To identify the relationship between genders satisfaction level from the food and beverage outlets with overall holiday satisfaction, Spearman's rank correlation coefficient test was applied. In the results of test, r value was calculated at 0.337 without consideration of gender. When the gender was considered, r value was calculated at 0.408. This confirms that there is relationship between gender's food and beverage department between satisfaction and overall satisfaction.

*H3.1: There is relationship between the education levels of tourist with comparison the all-inclusive hotels quality to standalone 5 stars hotels quality.*

Results show the cross tabulation and chi-square test results between education and comparison of all inclusive resort hotels with standalone 5 star hotels. We can say that there is relationship between the education levels of tourist with comparison the all-inclusive hotels to standalone 5 star hotels. As the education level increased, satisfaction level from all-inclusive hotels increased, which is very interesting. Also another not expected outcome was that once comparisons are done by tourist, it can be observed that 5 star all inclusive hotels are perceived to serve higher quality than standalone 5 star hotels. This however is an indication that needs further research in other geographic areas and different clientele geographic distribution in order to have more conclusive results. It is however, a clear indication of shifting perceptions on all-inclusive establishments.

*H4.1: There is relationship between reasons to choose an all-inclusive resort hotels and involving any activity outside of hotel during the stay.*

This hypothesis was created to understand whether tourist stay in the hotel during their holiday or they do involve any activities outside of the hotel. Spearman's Chi-square test results show that there is significant statistically negative relationship between no obligations on tracing extra expenses and involving any activity outside of the hotel. When we do correlation analysis between involving in any activity outside the hotel with the reason to choose all-inclusive system is the relationship, the r value was -0.615. This indicates that there is a strong negative relationship between these two variables. Therefore we can conclude that people who never go out of the hotel are mainly preferred all-inclusive system the reason is clear they do not have to consider spending or tracing extra money. Additionally 39,7% of the survey respondents *never went out of the hot*el, which is also consistent with previously described criticism. The reasons however need to be further investigated since we are not clear on whether this relates to convenience or whether the establishments implicitly or explicitly discourage the tourists to leave the premises.

# 5 CONCLUSIONS

All-inclusive accommodation establishments and tour operators have become an indispensable element of a fast growing segment. All-inclusive package tours offer economical and hassle free holiday options, which is why it is often preferred by middle-income tourists and as this research illustrates now other profiles are joining the target market. Despite the heavy criticism on the negative impacts associated with all-inclusive resorts mostly with regards to the lack of recognition of local cultures and limited contribution to the development of local economies of the destinations, the segment seems to be set for further growth.

This study illustrates that, tourist consumer behaviour is generally changing due to convenience reason and this affects heavily the choice of holiday types as well as destinations. Tourists are now more price sensitive than ever and this is clearly the most important factor for holiday product type selection. In general, there has been a belief that all-inclusive system preferred by low-income tourists but the current analysis showed that income levels of tourists in all-inclusive resorts in Antalya higher than expected. This should be an interesting observation for hotel business pricing strategies as a means to influence on tourists' buying preferences. The quality of the hotel considered to be important factor when choosing an all-inclusive holiday by any type of tourist standards. Education level was not found to be important and interesting enough when the educational level of tourist increases the satisfaction degree obtained from an all-inclusive system is increasing, too. This could be due to excellent services or a broader perception of what constitutes a good deal. The destination's natural beauty, historical riches and cultural elements could be included in all-inclusive package tours but these research shows that the destination elements are less significant in the choice of holiday and this should be further investigated since it poses major issues to service providers and marketer who will find it rather difficult to identify differentiating features to compete with.

Concluding we feel that despite the criticism received, all-inclusive tourism products will only grow especially among Mediterranean countries mostly due to prevailing economic conditions and consequently shifting consumer patters. This will leave destinations, providers and marketers in a conundrum with regards to tourism development options as well as marketing promotion approaches. The segment of all-inclusive products is changing by becoming more sophisticated with better provisions and by attracting new types of tourists.

# 6 LIMITATIONS

The survey had been conducted solely on international tourist who stayed at 5 star all inclusive hotels in the province of Antalya. The province of Antalya has a big scale of resort area and definitely indicative but this research should be also open to other areas since this might change the synthesis of tourist and possibly the outcomes. Larger sample size would also be beneficial.

# REFERENCES

1. Aguilo, E and Rosselio, J (2012) Research note: The new all-inclusive board formula in mature destinations- from motivation to satisfaction, *Tourism Economics*, Vol 18 No 5.

2. Baloglu, S. and Uysal, M. (1996) *Market Segments of Push and Pull Motivations: A Canonical Correlation Approach*, International Journal of Contemporary Hospitality Management, 8 (3): 32-38.

3. Ciftçi,H., Düzakın, E., Önal, Y. (2007) *All Inclusive System and Its Effects on the Turkish Tourism*, Sector. Problems and Perspectives in Management / Volume 5, Issue 3.

4. Ernst and Young 2012, EMEIA_hotel_outlook_for_2012

5. http://www.ey.com/Publication/vwLUAssets/EMEIA_hotel_outloo k_for_2012/$FILE/Hotels_workshop_summary.pdf

6. Icoz, O. (2001). *Turizm İşletmelerinde Pazarlama*, (2.Baskı) Ankara: Turhan Kitabevi.

7. Irmak, T. (2003b). *Purpose of All-Inclusive*, Gündem Gazetesi, 8/8/2003, 6.

8. Henry, B (1989) Self-contained all-inclusive resort-hotels and small tourism business in Jamaica, *Tourism Review*, Vol. 44 Issue: 2, pp.17 – 18

9. Hoffman and Bateson (2011), *Services Marketing: Concepts, Strategies & Cases*, 4th International edition, London: Thompson South Western.

10. Hotel Price Index (2013)

11. http://www.hotel-price-index.com/2013/spring/global-price-change/index.html (last accessed, 21.06.2013)

12. John J. Issa, ChandanaJayawardena, (2003) The "all-inclusive" concept in the Caribbean, *International Journal of Contemporary Hospitality Management*, Vol. 15 Issue: 3 pp. 167 – 171.

13. Koca, E (2006) The implications of all-inclusive pricing system on internal and external customer satisfaction in the Turkish tourism market *Total Quality Management & Business Excellence* Volume 17, Issue 7.

14. Kotler, P. (2002) Marketing *Management*, The Millennium Edition. Pearson Custom Publishing.

15. Kotler, P., Bowen J. T. and Makens J. C. (2006) *Marketing for Hospitality and Tourism,* 4th ed. New Jersey: Pearson Education-Prentice Hall.

16. Kozak, M. (2001) Repeaters' behavior at two distinct destinations, *Annals of Tourism Research*, 28 (3), 784-807.

17. Kozak, M. (2002) Comparative Analysis of Tourist Motivations by Nationality and Destinations, *Tourism Management,* 23 (3): 221-232.

18. Kozak, N. (2006) *Tourism Marketing,* Ankara: Detay Publishing.

19. Laws, E. (1997) *Managing Packaged Tourism: Relationships, Responsibilities, and Service Quality in the Inclusive Holiday Industry,* London: International Thomson Business Press.

20. Lobato, L.H., Radilla, M.M.S., Tena, A.M. and Darcia, J.S. (2006) *Tourism Destination Image, Satisfaction and Loyalty: A Study in Ixtapa - Zihhuatanejo, Mexico,* Tourism Geographies, 8(4): 343-358.

21. Menekse, R. (2005) Evaluation of All inclusive System and Its Effects to Stakeholders at a Glance of the Hotel Managers *(The case of Marmaris), Journal of Economic and Social Research*, Spring 2005, 1:97-124.

22. Poon, A. (1998). All-Inclusive Resort, *Travel and Tourism Analyst,* (6): 62-77.

23. Voss, K. (2003). All Inclusive Is Not A Temporary Trend, *Resort Tourism and Travel Journal*, Vol. 10, pp. 14.

24. Yoon, Y., Uysal, M. (2005) An Examination of the Effects of Motivation and Satisfaction on Destination Loyalty: A Structural Model, *Tourism Management*, 26 (1): 45-56.

25. WTTC, http://www.wttc.org/site_media/uploads/downloads/Economic_Impact_of_TT_2013_Annual_Update_-_Summary.pdf, (last accessed, 21.06.2013)

6

# RETURNING FROM THE HORIZON: INTRODUCING *URBAN ISLAND STUDIES*

**Adam Grydehøj[1], Xavier Barceló Pinya[2], Gordon Cooke[3], Naciye Doratlı[4], Ahmed Elewa[5], Ilan Kelman[6], Jonathan Pugh[7], Lea Schick[8], R. Swaminathan[9]**

[1]Department of Human Geography, Lund University, Sweden
[2]Department of Catalan Philology & General Linguistics, Universitat de les Illes Balears, Spain
[3]Faculty of Business Administration, Memorial University of Newfoundland, Canada
[4]Department of Architecture, Eastern Mediterranean University, North Cyprus
[5]Department of Architecture, Helwan University, Egypt
[6]Institute for Risk and Disaster Reduction & Institute for Global Health, UniversityCollege London, United Kingdom & Norwegian Institute of International Affairs (NUPI)
[7]Department of Geography, Newcastle University, United Kingdom
[8]Technologies in Practice, IT University of Copenhagen, Denmark
[9]Azim Premji Philanthropic Initiatives, India

## ABSTRACT

Island studies tends to focus on peripheral, isolated, and marginal aspects of island communities, while urban studies has showed scant awareness of islandness: Although many people research cities on islands, there is little tradition of researching island cities or urban archipelagos *per se*. Island cities (densely populated small islands and population centres of larger islands and archipelagos) nevertheless play import cultural, economic, political, and environmental roles on local, regional, and global scales. Many major cities and ports have developed on small islands, and even villages can fulfil important urban functions on lightly populated islands. Island concepts are also deployed to metaphorically describe developments in urban space. The journal *Urban Island Studies* explores island and urban processes around the world, taking an island approach to urban research and an urban approach to island research.

## KEYWORDS

island cities, island studies, urbanisation, urban island studies, urban studies

## 1. THE SETTING

In his 2006 editorial introduction to the inaugural issue of *Island Studies Journal*, Godfrey Baldacchino (2006a: 3-5) set the stage for the field of island studies by presenting a list of facts on island population, government, economics – and research history. Long before island studies had been identified and established as a field, scholars were researching islands, and long before scholars were researching islands, people had assumed that islands were special (Gillis, 2007; Lowenthal, 2007), even if they could not always quite pinpoint why.

The emerging field of urban island studies has no such advantage. People tend not to connect the island with the urban. Island towns and cities *per se* are not generally regarded as special in any way. An island city might be special as a city *or*as an island – but not as an island city.

It might thus be worthwhile to provide a setting for urban island studies:

- Six of the world's ten most populous urban agglomerations (Brinkhoff, 2014) were established on small islands: Tokyo, Guangzhou, Jakarta, Manila, Mumbai, and Mexico City.
- The largest and most densely populated cities in the United States (New York City) and sub-Saharan Africa (Lagos) are based on small

islands, and the largest city in South America (São Paulo) is a historical offshoot of the island city of São Vicente.

- European capital cities based or founded on small islands include Amsterdam, Copenhagen, London, Paris, and Stockholm. In addition, Europe's small island states and substantially autonomous subnational island jurisdictions include the capital cities of Douglas, Isle of Man; Mariehamn, Åland; Nicosia, Cyprus; Nuuk, Greenland; Reykjavík, Iceland; Saint Hellier, Jersey; St Peter Port, Guernsey; Tórshavn, Faroe; and Valletta, Malta.

- The arc of East Asian, Southeast Asian, and Pacific island states (Japan, Taiwan, Indonesia, Philippines, Singapore, Indonesia, and Papua New Guinea) have a collective population of 515,120,000.

- Fifteen of the world's twenty most densely populated sovereign states and substantially autonomous subnational jurisdictions are wholly or substantially based on small islands: Macau, Singapore, Hong Kong, Bahrain, Malta, Bermuda, Sint Maarten, Maldives, Jersey, Guernsey, Saint Martin, Taiwan, Barbados, Mauritius, and Mayotte.

- In the Middle Ages, small island trading centres played key roles in the movement of products, people, and ideas between East and West, North and South: Bruges, Gotland, Lamu, Lübeck, Mozambique Island, Ormus/Hormuz, Pemba, Venice, Zanzibar, etc.

- Eight of the world's ten busiest ports (by volume) are based on archipelagos: Shanghai, Singapore, Tiajin, Rotterdam, Guangzhou, Ningbo-Zhoushan, Busan, and Hong Kong.

These figures, of course, merely tell us that certain patterns of human settlement and economic activity frequently take place on islands. They do not explain why islands might tend to be densely populated, why they might encourage the formation of major cities, why they might be significant players in the global exchange of goods and ideas, how they might host particular kinds of cultural processes, and why it might be interesting to examine certain cities as island cities*per se*. The field of urban island studies seeks to answer these and other questions. In so doing, urban island studies will contribute to our knowledge concerning islands and cities more broadly.

# 2. ISLAND STUDIES

The field of island studies is often defined as the study of islands 'on their own terms' (McCall, 1994; Baldacchino, 2004). This definition is so attractive in part due to its implied resistance to the study of islands on *someone else's* terms. It is a rallying cry, a protest against the injustice of the mainlanded many taking it upon themselves to define and research the islanded few.

Islands have for so long and so often functioned as research sites and units of artistic currency – as microcosms in which to test macrocosmic hypotheses – that the "the lure of the island" appears almost irresistible (Baldacchino, 2012, 2006a). The field of island studies does not deny the importance of islands to mainlanders, yet it stresses the primacy of islanders' perspectives. Hau'ofa (1994: 152) calls for a move from "viewing the Pacific as 'islands in a far sea'" to viewing it as "'a sea of islands'": It is necessary to place islands – and their surrounding seas – at the centre instead of marginalising them as "dry surfaces in a vast ocean far from the centers of power." Although many small island societies with distinctive cultures, politics, and economies may indeed be "empire dust" (Couroucli, 2010), remnants of geopolitics gone by, they are also places in their own right.

It is on these foundations that the field of island studies has been built. Researchers from numerous disciplines have sought to realign island scholarship, to transform 'the island' from a mere locus of research to a focus of study (Ronström, 2012):

- Island politics and government: Researchers have emphasised small island societies as centres of power and political innovation as well as enmeshed in global politics (e.g. Anckar, 2007; Baldacchino, 2014, 2010; Baldacchino & Milne, 2000; Baldacchino & Tsai, 2014; Bartmann, 2006; Coddington *et al.*, 2012; Grydehøj, 2013; Grydehøj *et al.*, 2012; Karlsson, 2009; Loyd & Mountz, 2014; Mountz, 2014).
- Island economies: Researchers have explored spatial and political contexts to understand the workings of a specifically insular economics (e.g. Baldacchino, 2006b; Bertram & Watters, 1985; Briguglio *et al.*, 2010; Connell, 2013; Grydehøj & Hayward, 2014;

Grydehøj & Nurdin, 2015; Kakazu, 2012; Karides, 2013; McElroy *et al.*, 2014; Tsai & Chiang, 2014).

- Island business (including tourism): Researchers have contributed to a complex picture of how physically isolated communities are drawn into the global market as well as the hidden costs and benefits of this movement (e.g. Baldacchino & Ferreira, 2013; Baum *et al.*, 2000; Cooke *et al.*, 2013; Hampton & Christensen, 2011; Hayward, 2014; Johnson, 2012; Ioannides*et al.*, 2001; Royle, 2009).

- Island heritage, culture, and tradition: Researchers have revealed how island communities accrue influences from and influence the wider world as well as how feelings of culture, ethnicity, and nationality affect island politics and economics (e.g. Grydehøj & Hayward, 2011; Guerrero, 2013; Johnson & Kuwahara, 2013; McCusker & Soares, 2011; Ronström, 2008; Tabucanon, 2014).

- Environment: Researchers have studied human-environment interactions as well as environmental risks, adaptation, resilience, and sustainability (e.g. Christensen & Mertz, 2010; Mertz *et al.*, 2010; Hayward & Mosse, 2012; Kelman *et al.* 2012, 2011; Kelman & West, 2009; Maceda *et al.*, 2009; McIntyre, 2012; Stratford *et al.*, 2013).

Much of the above social sciences research is strongly influenced by disciplinary traditions from geography. Indeed, this geographic orientation has sometimes resulted in the sidelining of approaches from literary studies and related disciplines (Fletcher, 2011). Nevertheless, recent years have seen island studies become more solidly theorised, with writers like Baldacchino (2008; Baldacchino & Kelman, 2014), Clark (2013), Depretaere (2008a, 2008b), Hay (2013, 2006), Hayward (2012a), Pugh (2013a), and Stratford (2013; Stratford *et al.*, 2011) nuancing the field's approach and subject matter as well as problematising the field's focus on island remoteness, isolation, and peripherality.

As island studies has grown in sophistication, so has the field grown in its capacity to communicate research. In terms of peer-reviewed journals, *Island Studies Journal* was founded in 2006, *Shima: The International Journal of Research into Island Cultures* dates from 2007, and *Journal of Marine and Island Cultures* was established in 2012. A number of organisations also run regular series of interdisciplinary, international conferences, including: International Small Islands Studies Association, Small Island Cultures Research Initiative, International Geographical Union's Commission on Islands, and Island Dynamics.

Island studies thus seems to be on the cusp of transitioning from an emerging field into an established field – if this transition has not already occurred. Until recently, however, the scope of island studies, as it has developed over the past decades, has arguably not fully accounted for or acknowledged one key category of island places: island cities. Island cities are the focus of the new journal *Urban Island Studies*.

Many island studies researchers do, of course, study cities and densely populated islands. It is just that the field has tended not to cast such places as urban or as cities (exceptions include Hayward, 2012b; Phua, 2014). As illustrated above, there are in fact a great number of people packed onto a great number of often-densely populated islands. As a result, island studies' emphasis on 'the islanded few' does not merely protest against how peripheral island communities are denied centrality; it also denies island cities true island status. Ironically, island studies has failed precisely in presenting island communities as centres: Island cities are regarded as not truly insular, and peripherality becomes a defining feature of islands. The terms of the debate are skewed against understanding islands on their own terms. In such a discourse, we cannot hope to place island communities at the centre; we can at best succeed in presenting islands as self-centred.

The *Urban Island Studies* ethos represents a return from the island studies horizon. It seeks to fulfil island studies' promise of understanding islands on their own terms precisely by combatting the futile provincialisations of an island rhetoric that tacitly accepts the dominance of the centre in its championing of the periphery. Only by permitting ourselves to regard islands as centres within complex networks of centrality and peripherality – rather than simply as peripheries or as mystical centres unto themselves – can we transcend the savage inequalities of how we speak about and act upon both islands and mainlands. Only then can we place islands in context and place context on islands.

# 3. TYPES OF ISLAND CITIES

We can break geographical island cities down into two primary categories: 1) major population centres of large islands or archipelagos and 2) densely urbanised small islands or archipelagos. We recognise that this categorisation is not absolute and that other categories and subcategories are feasible, but we put forward this bifurcation as a starting point for discussion and development.

For major population centres of large islands or archipelagos, it is not the city's absolute size (in terms of population, economy, coastline length, land area, or water area/volume under its jurisdiction) that is important but instead its relative size (compared with the surrounding area) and/or its fulfilment of urban functions. It is thus that, while Taipei and Manila are very large cities on very large and highly populous islands and archipelagos, small places like Longyearbyen, Svalbard (population 2000) and Hugh Town, Isles of Scilly (population 1100) are likewise island cities inasmuch as they perform urban functions and are population centres for small and lightly populated archipelagos. Some extremely remote island communities – such as the UK territories of Pitcairn and Tristan da Cunha – are only accessible by boat following a journey of several days since they tend to be out of float plane and helicopter range. Even some islands that are not quite so distant – such as peripheral islands in some of the vast Pacific archipelagos – may in practice be accessible solely by infrequent and/or lengthy transport. Such remote islands can possess a great degree of self-dependency and either formal or informal administrative autonomy so that even the smallest of settlements can take on the functions of much larger territories (Amoamo, 2013; Royle, 2010). Furthermore, elements of urban design, architecture, and planning apply even to tiny and rural island communities.

Although some island communities remain truly remote, improved transport technologies and infrastructures (high-speed boats, bridges, airplanes, tunnels, causeways, cars, motorways, railways, etc.) have made many formerly difficult-to-reach small island communities relatively easily accessible. One might imagine that globalisation would decrease the significance of the island cities on which these remote communities depend, but it has in fact reinforced the centrality of the old island cities: The more complex and resource-heavy the transport technology, the more important the island city's hub and gateway functions relative to its periphery. It is thus, for example, that even though (or perhaps because) the Penghu archipelago is now just a short flight from Taipei on mainland Taiwan, Penghu's main town of Magong has continued accruing significance as the islands' transport hub and business centre. Similarly, the tyranny of rising expectations means that the remote settlements of East Greenland are now dependent on West Greenland's transport hub (Kangerlussuaq) and administrative, industrial, and business centre (Nuuk) – which are themselves dependent on the distant island city of Copenhagen. Understanding the interconnected urban functions of island cities as diverse as Paris and Saint-Denis de la Réunion, as Reykjavík and Akureyri, as Jakarta and Ambon is vital for understanding life in general on large islands and archipelagos. There is furthermore substantial opportunity to explore the causes and effects of employment, economic, cultural, and other differences between those living within these island cities and those in rural and/or remote locations.

While mainstream island studies has often considered urban centres of large archipelagos (though rarely from an explicitly urban perspective), the second category of island city (densely urbanised small islands or archipelagos) has received less attention. This category can be further divided into four (overlapping) subcategories:

(a) Cities that are contiguous (or nearly contiguous) with one or more small islands (Abu Dhabi, Malé, Miami Beach, Mumbai, Singapore, etc.),

(b) Cities that are substantially or significantly located on one or more densely urbanised small islands (Florianópolis, Guangzhou, Lagos, Montreal, Stockholm, etc.),

(c) Small islands within cities that are largely located on the mainland (the islands of Cairo, Ho Chi Min City, Shanghai, Tokyo, Toronto, etc.),

(d) Small islands or archipelagos that cannot be regarded as a single urban zone but that are nonetheless densely urbanised as a whole and/or fulfil urban functions (Bahrain, Barbados, Malta, Penang, Zanzibar, etc.).

Many cities in subcategory (b) began as settlements based on small islands or archipelagos but have since expanded outwards onto the mainland (Hong Kong, Mombasa, São Vicente, St Petersburg, etc.). In other cases, land reclamation processes have led to islands physically joining the mainland (London, Macau, Manhattan, Mumbai, etc.). This is not to mention the presence of bridges, causeways, and tunnels (fixed links), which connect many densely urbanised small islands to adjacent landmasses. Although it might appear odd to regard cities that are no longer limited to islands as 'island cities', the historical topography and island spatiality of these cities has often strongly conditioned their future development. 'Dry' and linked islands must be included in our definition of islands if we are to avoid artificially isolating and peripheralising island society from mainland society (Barrowclough, 2010).

Just as mainlands are swallowing up islands through land reclamation, they are also producing new islands and expanding old ones. The Palm Islands and World archipelago of Dubai might be the most famous examples of artificial islands at present, but island spaces of cities around the globe – from Manhattan to Bahrain, from Singapore to St Petersburg, from Amsterdam to Osaka – have been created and altered through human engineering.

Besides the two above categories of island cities, we can also note the frequent metaphorical uses of islands and archipelagos within urban studies, architecture, and planning. That is to say, certain strands of urban research and thought find it useful to deploy island terminology as a means of understanding how urban areas, settlements, buildings, infrastructure, and so on function or ought to function. Such metaphorical approaches (considered below) must also fall within the scope of urban island studies as instances of the intersection of social and geographical understandings of island spatiality within an urban context.

## 4. URBAN STUDIES

Urban studies is a well-established research field, with a range of top-quality journals and a history of making an impact on other fields. Prestigious periodicals include *Cities*; *City & Community*; *City, Culture and Society*; *Environment and Urbanization*; *Environment and Planning B: Planning and Design*; *European Urban and Regional Studies*; *International Journal of Urban and Regional Research*; *Journal of Urban Affairs*; *Journal of Urban Design*; *Journal of Urban Economics*; *Landscape and Urban Planning*; *Regional Science and Urban Economics*; *Regional Studies*; *Territory, Politics, Governance*; *Urban Affairs Review*; *Urban Geography*; and *Urban Studies*.

Island studies has perhaps given island cities less attention than they deserve in part because the dominant discourse within the field – laudably seeking to de-marginalise remote communities – might seem initially to make little sense when applied to the metropolis, even though many locations within cities are marginalised, peripheralised, islanded, and separated from the core or the power. It is more difficult to explain why urban studies has similarly shown little awareness of islands, for as demonstrated above, cities on islands play disproportionately prominent roles on the global stage, and many island aspects and traits emerge in urban environments. The lack of awareness of islands within urban studies may partially come down to the landscape changes prompted by urbanisation processes. Small island status has been made invisible by centuries of land reclamation in Manila, Mexico City, and Tokyo as well as by the outward spread of urbanisation from historical small island urban centres like Lagos, Paris, and Rio de Janeiro. By the same token, the ease of present-day mobility makes us forget past limits to mobility: Transport technologies and infrastructure close the gaps between the multitudes of islands and peninsulas in urban archipelagos like Busan, Hong Kong, and Stockholm – not to mention between island and mainland states.

Urban studies is largely informed by structural approaches that reductively imagine the city and its urban spaces – its numerous urbanisms so to speak – as sociocultural, economic, and political specificities of place rather than as historiographies of processes with material and physical foundations. In this sense, the city itself becomes problematic. One issue is the manner in which the dominant theoretical frameworks of study (modernity, urbanity, colonialism, postcolonial cityscapes, technosocial global cities, etc.) are used as modular structural scaffolding. This creates unintentional hierarchy – and a widening gap – between urban theory and the organically produced multivocal and intersected narratives of the city. By default, the conventional approaches first position the theoretical rubric and then seek to fit the city's experiences within it as one would the pieces of a puzzle.

Another issue involves the marginalisation of the city's roots in the creation of its stories, the manner in which the organic narratives of local and indigenous communities have been subsumed by dominant storylines (Swaminathan, 2015). There is also a near-complete absence of serious scholarship even acknowledging – let alone studying and analysing – the archipelagic, coastal, and littoral nature of the island city as valid material and non-material foundations for the city's territorial, spatial, and cultural construction.

The field of urban studies is, however, undergoing change. Recent decades have seen several fruitful attempts to introduce more qualitative socio-anthropological and sociocultural approaches to the field (e.g. Appadurai, 2006, 2001, 1986; Mayaram, 2012). Such studies have introduced a certain 'human quality' to the city, regarding urban spaces as organic entities. This has expanded the landscape of urban studies, producing new methodological tools and lines of inquiry that can be usefully applied to island cases. Island studies has thus far been slow to interact with and take on board urban-inspired theories of governance and territory (e.g. Brenner, 2004; Pierre, 2011; Sassen, 2013, 2006; Uitermark, 2015). Island studies could likewise benefit from nuanced urban explorations of the interactions between urban design, infrastructure, resources, space, and society (e.g. Elewa & El-Garhy, 2013; Graham & Hewitt, 2013; Graham & Marvin, 2001; Kaika & Swyngedouw, 2000; Kong & Yeoh, 2003; Pasaogullari & Doratli, 2004; Tonkiss, 2013), which may prove applicable to island contexts of all sorts.

More challenging for island studies are the arguments that the world has already been completely urbanised and that the 'traditional' city has been replaced by global urban processes (e.g. Brenner, 2014, 2013; Wachsmuth, 2014; Amin & Thrift, 2002), with the result that 'the urban' now encompasses not just the usual suspects (like New York City and Mumbai) but also remote and sparsely inhabited islands (like Greenland and Kiribati). Indeed, small island spatiality may encourage exceptionally comprehensive urbanisation even in communities typically regarded as extremely peripheral, highlighting the need for improved urban governance and understanding of the drivers and impacts of urban development (Barnett & Margetts, 2013; Bautista, 2011; Bryant-Tokalau, 2014; Jones, 2012; Reis & Hayward, 2013).

This is not to say that the urban has been completely forgotten when it comes to the study of islands, especially in regions such as the Caribbean, where scholars have for many years researched densely populated urban coastal corridors (Pugh & Potter, 2003; Clark & Howard, 2006). Yet as Hay (2013) points out concerning Caribbean studies more generally, there has been a tendency to downplay the*islandness* of the region and the role of the sea in particular. Such concerns bring to the fore questions about the appropriateness of 'imported' research techniques, models of island urban planning, island/sea boundaries, and the recurrent theme of what it means to expand and develop the power of island voices as well as what it means to discover their terms (Pugh, 2013, 2005; Sletto, 2002).

As noted above, urban research has also used islands metaphorically. Islands have served as models and metaphors for cities since ancient times (Pigou-Dennis & Grydehøj, 2014), but it was the manifesto on *The*

*City in the City – Berlin: The Green Archipelago* by the architects Ungers and Koolhaas (2013 [1977]) that helped revive the popularity of islands within modern urban design. For Ungers, Koolhaas, and their collaborators, the depopulating European city could ideally evolve into an 'archipelago city' (*Stadtarchipel*) consisting of islands of urban density within a sea of forest (Neumayer, 1990). This politico-architectural vision of positive urban polycentricity has continued to attract interest (e.g. Aureli, 2011; Schrijver, 2006). It has also contributed to a distinct strand of island terminology emphasising the negatively fragmented and militarised nature of the postmodern city. Drawing upon Foucault's nightmare of the 'carceral archipelago' (Foucault, 1999 [1975]), island metaphors are deployed to examine processes of urban social and economic exclusion: "islands of enclosure and anticipated protection against the real and imagined dangers of daily life" (Soja, 2000: 299); "the rise of residential archipelagos that are driven by aspects of security and exclusiveness" (Borsdorf *et al.*, 2007: 367); "security strategies turn open city districts into temporary, mobile 'security islands' replete with paramilitary forces, militarized cordons, and even surface-to-air missiles" (Graham, 2010: 121).

Problematically, both these positive and negative deployments of the island as an urban metaphor make isolation the essence of islandness. This evokes the stereotype of islands as places essentially disconnected, cut-off, and apart. Unlike the stereotype, however, island communities tend overwhelmingly to be integrated into complex networks of exchange for goods, people, and ideas. An important task for future research within urban island studies must be to nuance island metaphors and assist in bringing knowledge of real-world island dynamics to bear upon analysis of island-like processes in urban space. The island has been so attractive as a metaphor over the course of history that it would be futile to deny its metaphorical value; we should instead engage with the metaphorical uses of islands to ensure that they reflect their desired content and do not feed into a broader peripheralising discourse. There is a valid and strong case for the archipelagic lens to become not only an organic frame of reference but also a theoretical structure for analysing the city. The lens is robust enough to integrate into dominant theoretical frameworks, potentially bringing much-needed sociocultural context and organic flexibility into the existing approaches to urban research.

# 5. URBAN ISLAND STUDIES: THE FIELD OF STUDY

The field of urban island studies had its breakthrough in 2014 due to activities coordinated by the Island Dynamics organisation. Island Dynamics produced a special thematic section of *Island Studies Journal* (Vol. 9, No. 2, pp. 183-292) on the theme of island cities. This special section included an editorial introduction by Adam Grydehøj (2014b), plus six papers on a range of urban island studies topics with geographic focus on Greenland, the Balearic Islands, Belize City, Mumbai, and Isle of Man (Grydehøj, 2014a; Picornell, 2014; Pigou-Dennis & Grydehøj, 2014; Pons *et al.*, 2014; Swaminathan, 2014; Tutt, 2014). Although these papers discussed disparate topics, they together made a case for an urban island studies that can both take in the big picture of big places and be sensitive to the shifting livelihoods, lifestyles, and landscapes that see communities transition from urban to rural and back again. The papers also, however, highlighted just how little research has thus far been performed on how islandness affects urban development and how urban development affects islands, thereby emphasising the need for additional study.

The *Island Studies Journal* special section was published in October 2014 to coincide with the inaugural Island Cities and Urban Archipelagos (ICUA) conference, held in Copenhagen. The conference featured representation from a wide range of academic fields and disciplines, including human geography, urban planning and design, architecture, literature, heritage studies, business studies, transport, disaster resilience (including climate change), tourism studies, history, and archaeology. The ICUA conference also hosted the launching of the Island Cities and Urban Archipelagos research network (http://www.islandcities.org) under the Island Dynamics umbrella. The research network will organise future ICUA conferences, publications, and research projects. The ICUA 2016 conference will be held in Hong Kong, further establishing and strengthening urban island studies as well as providing a platform for communicating research from the field.

Urban island studies offers the potential to improve our understanding of both urban and island development processes. The fact, for instance, that strategically located small islands are so disproportionately likely to host major urban centres suggests that research into the islandness of such island cities could tell us something about cities in a broader sense. Because cities on small islands tend to be exceptionally densely urbanised relative to other spaces, the extreme citynesss of island cities may make urban processes more distinct, thereby aiding in their study. Similarly, knowledge of centre-periphery relationships within larger islands and archipelagos may clarify the dynamics of such relationships more generally for the precise reasons why islands are often otherwise selected as laboratories and research sites.

Taking an urban approach is also important for island studies. As a field, island studies has been unable to convincingly identify what, if any, effects islandness has on societal development. This is no surprise given that cities have been largely lacking from island studies analyses, rendering the field's understanding of the nature of islandness incomplete. However, by researching how island geographies influence urban areas, the field could more easily differentiate between effects of islandness and effects of small population and/or economic size (Grydehøj, 2014b: 187). Perhaps more significantly, an urban perspective is necessary to add nuance to reigning island studies understandings of island societies. An island studies that truly studied islands on their own terms would not persist, for example, in analysing towns like Papeete, French Polynesia (population 25,800) and Kirkwall, Orkney (population 9300) solely as remote communities when these are, in fact, the major population, industrial, and transport centres of their respective archipelagos. It is important to seek to understand the urban functions filled by such island cities, regardless of their size relative to mainland cities. This should furthermore extend to studies of 'second cities', 'third cities', and beyond in order to analyse the full complexity of urban-rural and inter-urban relationships within an island jurisdiction or physical delineation. In the words of Baldacchino (2006a: 10):

'Island studies' need/should not be focused only on islands themselves, but also on relations between islands and mainlands. [...] Seeing islands as part of complex and cross-cutting systems of regional and global interaction should be one of the strengths of island studies as well as of an island studies journal.

This should also be a strength of an urban island studies journal.

## 6. *URBAN ISLAND STUDIES*: THE JOURNAL

It is in this context that the *Urban Island Studies* journal has been created. *Urban Island Studies* is an online, open access journal, harnessing the internet to bypass traditional publishing industry business models, which frequently serve to further entrench existing power structures. All research papers published in this journal are subject to double-blind peer review by two or more reviewers. Papers are published on a rolling basis and assembled into annual volumes.

*Urban Island Studies* publishes in-depth case studies and internationally comparative papers, which can add to the empirical basis of the field, as well as theoretical contributions. Papers will consider historical or present-day cases, including islands that are now part of the mainland. The journal's scope includes but is not limited to research articles concerning:

- Densely urbanised small islands,
- Urban centres of larger islands or archipelagos,
- Islands within mainland cities,
- Urbanisation and urban processes in island communities,
- Relationships between urban centres and peripheral island communities,
- Metaphorical uses of islands and archipelagos in urban research,
- Papers contributing to urban island studies' theoretical foundations.

Research is welcome that is theoretical, empirical, policy orientated, or combines these approaches. With this in mind, the journal will engage with such questions as:

- How does island spatiality affect urbanisation processes?
- Are island cities subject to certain kinds of cultural dynamics as a consequence of urban density and hub functions?
- How can island residents and governments adapt to environmental risks prompted by urbanisation?
- What roles do island cities play in global networks (cultural, political, and economic)?
- Do islands host particular problems with regard to urban poverty and wealth disparity?

- How does land scarcity influence patterns of land use, land change, and land ownership?
- Does island spatiality encourage the development of certain architectural and urban design solutions?
- How does island urbanisation affect employment and human migration processes?
- What special challenges does islandness present for urban sustainability?
- How can the island metaphor help us understand urban life in general?
- How can island cities exploit their status to compete in the global economy?
- How does urbanisation affect island transport infrastructures and island spatiality itself?
- How do rural island communities interact with their (island and mainland) urban centres?
- How does island city urban governance differ from that of mainland cities?

These are just a few of the many topics that will be discussed in *Urban Island Studies*. These various inquiries are tied together by an interest in how the island and the urban interact. In order to avoid excluding those from policy and practitioner backgrounds, authors who might be unaccustomed to the format of academic publishing are welcome to work with an Editorial Board member to learn expectations of academic writing. In order to maximise research accessibility,*Urban Island Studies* will also publish multilingual articles when requested, presenting versions of papers in non-English language(s) alongside the English-language version of record. Finally, non-traditional forms of presenting research (such as photo essays, videos with original choreography, or musical compositions) for peer-review might be considered.

Reflecting the fields of urban studies and island studies themselves, *Urban Island Studies* is comprehensively interdisciplinary. The journal's founding editorial board includes researchers with backgrounds in architecture; business and employment; disaster, resilience, and sustainability (including climate change); engineering; folklore; literature and language; participatory planning and development; political economy and urban affairs; sustainable cities, art, and science and technology studies (STS); and urban design.

## 7. CONCLUSION

*Urban Island Studies* will make valuable contributions to urban studies and island studies, especially by bringing the fields together. In time, urban island studies will become an accepted subfield of both these parent fields. *Urban Island Studies* is neither an island studies journal nor an urban studies journal but is instead an equal meeting of two vibrant scholarly traditions, both of which, however, have traditionally possessed certain limitations – as is true of all fields and as will certainly become true for urban island studies as well. By seeking to overcome these limitations – to bolster the urban within the island and the island within the urban – *Urban Island Studies* will help advance understandings of islands, cities, and everything in between.

## REFERENCES

1. Amin, A., & N. Thrift (2002) *Cities: Reimagining the Urban.* Cambridge: Polity.

2. Amoamo, M. (2013) 'Empire and Erasure: A Case Study of Pitcairn Island',*Island Studies Journal*, 8(2), 233-254.

3. Anckar, D. (2007) 'Archipelagos and Political Engineering: The Impact of non-Contiguity on Devolution in Small States', *Island Studies Journal*, 2(2), 193-208.

4. Appadurai, A. (2006) *Fear of Small Numbers.* Durham, NC: Duke University Press.

5. Appadurai, A. (ed.) (2001) *Globalisation*. Durham, NC: Duke University Press.

6. Appadurai, A. (1986) *The Social Life of Things: Commodities in Cultural Perspective*. Cambridge: Cambridge University Press.

7. Aureli, P.V. (2011) *The Possibility of an Absolute Architecture*. Cambridge, MA: MIT Press.

8. Baldacchino, G. (2014) 'Islands and the Offshoring Possibilities and Strategies of Contemporary States: Insights on/for the Migration Phenomenon on Europe's Southern Flank', *Island Studies Journal*, 9(1), 57-68.

9. Baldacchino, G. (2012) 'The Lure of the Island: A Spatial Analysis of Power Relations', *Journal of Marine and Island Cultures*, 1(2), 55-62.

10. Baldacchino, G. (2010) *Island Enclaves: Offshoring Strategies, Creative Governance, and Subnational Island Jurisdictions*. Montreal & Kingston: McGill-Queen's University Press.

11. Baldacchino, G. (2008) 'Studying Islands: On whose Terms? Some Epistemological and Methodological Challenges to the Pursuit of Island Studies',*Island Studies Journal*, 3(1), 37-56.

12. Baldacchino, G. (2006a) 'Islands, Island Studies, Island Studies Journal', *Island Studies Journal*, 1(1), 3-18.

13. Baldacchino, G. (2006b) 'Managing the Hinterland Beyond: Two Ideal-Type Strategies of Economic Development for Small Island Territories', *Asia Pacific Viewpoint*, 47(1), 45-60.

14. Baldacchino, G. (2004) 'The Coming of Age of Island Studies', *Tijdschrift voor economische en sociale geografie*, 95(3), 272-283.

15. Baldacchino, G., & E.C.D. Ferreira (2013) 'Competing Notions of Diversity in Archipelago Tourism: Transport Logistics, Official Rhetoric and Inter-Island Rivalry in the Azores', *Island Studies Journal*, 8(1), 84-104.

16. Baldacchino, G., & I. Kelman (2014) 'Critiquing the Pursuit of Island Sustainability: Blue and Green, with Hardly a Colour in Between', *Shima*, 8(2), 1-21.

17. Baldacchino, G., & D. Milne (eds) (2000) *Lessons from the Political Economy of Small Islands: The Resourcefulness of Jurisdiction*. Basingstoke: Macmillan.

18. Baldacchino, G., & Tsai H.M. (2014) 'Contested Enclave Metageographies: The Offshore Islands of Taiwan', *Political Geography*, 40, 13-24.

19. Barnett, R., & J. Margetts (2013) 'Disturbanism in the South Pacific: Disturbance Ecology as a Basis for Urban Resilience in Small Island States' in S.T.A. Pickett *et al.* (eds) *Resilience in Ecology and Urban* Design. Dordrecht & New York: Springer, pp. 443-459.

20. Barrowclough, D.A. (2010) 'Expanding the Horizons of Island Archaeology',*Shima*, 4(1), 27-46.

21. Bartmann, B. (2006) 'In or Out: Sub-National Island Jurisdictions and the Antechamber of Para-Diplomacy', *The Round Table*, 95(386), 541-559.

22. Baum, T.G., *et al.* (2000) 'Tourism and Cold Water Islands in the North Atlantic' in G. Baldacchino & D. Milnes (eds) *Lessons from the Political Economy of Small Islands: The Resourcefulness of Jurisdiction*. Houndsmills *et al.* & New York *et al.*: Macmilla & St Martin's, pp. 214-229.

23. Bautista, L.Q. (2011) 'Building Sense out of Households: Migrants from Chuuk (Re)create Local Settlements in Guam', *City & Society*, 23(1), 66-90.

24. Bertram, G., & R.F. Watters (1985) 'The MIRAB Economy in South Pacific Microstates', *Pacific Viewpoint*, 26(3), 497-519.

25. Borsdorf, A., *et al.* (2007) 'A New Model of Urban Development in Latin America: The Gated Communities and Fenced Cities in the

Metropolitan Areas of Santiago de Chile and Valparaíso', *Cities*, 24(5), 365-378.

26. Brenner, N. (ed.) (2014) *Implosions / Explosions: Towards a Study of Planetary Urbanization*. Berlin: Jovis.

27. Brenner, N. (2013) 'Theses on Urbanization', *Public Culture*, 25(1 69), 85-114.

28. Brenner, N. (2004) *New State Spaces: Urban Governance and the Rescaling of Statehood*. Oxford: Oxford University Press.

29. Briguglio, L., *et al.* (eds) (2010) *Profiling Vulnerability and Resilience: A Manual for Small States*. London: Commonwealth Secretariat.

30. Brinkhoff, T. (2014) 'Major agglomerations of the world'. Available at: http://www.citypopulation.de/world/Agglomerations.html. Accessed 17 December 2014.

31. Bryant-Tokalau, J.J. (2014) 'Urban Squatters and the Poor in Fiji: Issues of Land and Investment in Coastal Areas', *Assia Pacific Viewpoint*, 55(1), 54-66.

32. Christensen, A.E., & O. Mertz (2010) 'Researching Pacific Island Livelihoods: Mobility, Natural Resource Management and Nissology', *Asia Pacific Viewpoint*, 51(3), 278-287.

33. Clark, E. (2013) 'Financialization, Sustainability and the Right to the Island: A Critique of Acronym Models of Island Development', *Journal of Marine and Island Cultures*, 2(2), 128-136.

34. Clarke, C., & D. Howard (2006) 'Contradictory Socio-Economic Consequences of Structural Adjustment in Kingston, Jamaica', *The Geographical Journal*, 172(2), 106-129.

35. Coddington, K., *et al.* (2012) 'Embodied Possibilities, Sovereign Geographies and Island Detention: Negotiating the 'Right to have Rights' on Guam, Lampedusa and Christmas Island', *Shima*, 6(2), 27-48.

36. Connell, J. (2013) *Islands at Risk?: Environments, Economies and Contemporary Change*. Cheltenham: Edward Elgar.

37. Cooke, G.B., *et al.* (2013) 'The Nuanced Nature of Work Quality: Evidence from Rural Newfoundland and Ireland', *Human Relations*, 66(4), 503-527.

38. Couroucli, M. (2010) 'Empire Dust', in C. Hann & H. Goltz (eds) *Eastern Christians in Anthropological Perspectives*. Berkeley: University of California Press, pp. 220-239.

39. Depraetere, C. (2008a) 'The Challenge of Nissology: A Global Outlook on the World Archipelago. Part I: Scene Setting the World Archipelago', *Island Studies Journal*, 3(1), 3-16.

40. Depraetere, C. (2008b) 'The Challenge of Nissology: A Global Outlook on the World Archipelago. Part II: The Global and Scientific Vocation of Nissology', *Island Studies Journal*, 3(1), 17-36.

41. Elewa, A.K.A., & W.A.T. El-Gahry (2013) 'The Roles of the Urban Spatial Structure of the Main Cities Slums in Egypt and the Environmental Pollution' in S. Rauch *et al.* (eds) *Urban Environment: Proceedings of the 11th Urban Environment Symposium (UES), held in Karlsruhe, Germany, 16-19 September 2012, Part III*, Dordecht & New York: Springer, pp. 179-191.

42. Fletcher, L. (2010) '"... Some Distance to Go": A Critical Survey of Island Studies', *New Literatures Review*, 47-48, 17-34.

43. Foucault, M. (1995 [1975]) *Discipline and Punish: The Birth of the Prison*. A. Sheridan (trans.). New York & Toronto: Random House.

44. Gillis, J.R. (2007) 'Island Sojourns', *Geographical Review*, 97(2), 274-287.

45. Graham, S. (2010) *Cities under Siege: The New Military Urbanism*. London & Brooklyn: Verso.

46. Graham, S., & L. Hewitt (2013) 'Getting off the Ground: On the Politics of Urban Verticality', *Progress in Human Geography*, 37(1), 72-92.

47. Graham, S., & S. Marvin (2001) *Splintering Urbanism: Networked Infrastructures,*

48. *Technological Mobilities and the Urban Condition*. London: Routledge.

49. Grydehøj, A. (2014a) 'Constructing the Centre on the Periphery: Policies and Processes of Urban Development in Nuuk, Greenland', *Island Studies Journal*, 9(2), 205-222

50. Grydehøj, A. (2014b) 'Guest Editorial Introduction: Understanding Island Cities',*Island Studies Journal*, 9(2), 183-190.

51. Grydehøj, A. (2013) 'Challenges to Local Government Innovation: Legal and Institutional Impediments to the Exercise of Innovative Economic Development Policy by Subnational Jurisdictions, *European Journal of Spatial Development*, 50, 1-22.

52. Grydehøj, A., & P. Hayward (2014) 'Social and Economic Effects of Spatial Distribution in Island Communities: Comparing the Isles of Scilly and Isle of Wight, UK', *Journal of Marine and Island Cultures*, 3(1), 9-19.

53. Grydehøj, A., & P. Hayward (2011) 'Autonomy Initiatives and Quintessential Englishness on the Isle of Wight', *Island Studies Journal*, 6(2), 179-202.

54. Grydehøj, A., A. Grydehøj, & M. Ackrén (2012) 'The Globalization of the Arctic: Negotiating Sovereignty and Building Communities in Svalbard, Norway', *Island Studies Journal*, 7(1), 99-118.

55. Grydehøj, A., & N. Nurdin (2015) 'Politics of Technology in the Informal Governance of Destructive Fishing in Spermonde, Indonesia', *GeoJournal*, forthcoming.

56. Guerrero, P. (2013) 'A Story told through Plena: Claiming Identity and Cultural Autonomy in the Street Festivals of San Juan, Puerto Rico', *Island Studies Journal*, 8(1), 165-178.

57. Hampton, M.P., & J. Christensen (2011) 'Looking for Plan B: What Next for Island Hosts of Offshore Finance?', *The Round Table*, 100(413), 169-181.

58. Hau'ofa, E. (1994) 'A New Oceania: Rediscovering Our Sea of Islands', *The Contemporary Pacific*, 6(1), 148-161.

59. Hay, P. (2013) 'What the Sea Portends: A Reconsideration of Contested Island Tropes', *Island Studies Journal*, 8(2), 209-232.

60. Hay, P. (2006) 'A Phenomenology of Islands', *Island Studies Journal*, 1(1), 19-42.

61. Hayward, P. (2012a). 'The Constitution of Assemblages and the Aquapelagality of Haida Gwaii', *Shima*, *6*(2), 1-14.

62. Hayward, P. (2012b) 'Merlionicity: The Twenty First Century Elaboration of a Singaporean Symbol, *Journal of Marine and Island Cultures*, 1(2), 113-125.

63. Hayward, P., & J.W. Mosse (2012) 'The Dynamics and Sustainability of Ambon's Smoked Tuna Trade', *Journal of Marine and Island Cultures*, 1(1), 3-10.

64. Hayward, P., & G.T.H. Tran (2014) 'At the Edge: Heritage and Tourism Development in Vietnam's Con Dao Archipelago', *Journal of Marine and Island Cultures*, 3(2), 113-124.

65. Ioannides, D., *et al.* (2001) 'Searching for Sustainable Tourism Development in the Insular Mediterranean' in D. Ioannides *et al.* (eds) *Mediterranean Islands and Sustainable Tourism Development: Practises, Management and Policies.* London & New York: Continuum, pp. 3-22.

66. Johnson, H. (2012) "Genuine Jersey': Branding and Authenticity in a Small Island Culture', *Island Studies Journal*, 7(2), 235-258.

67. Johnson, H., & S. Kuwahara (2013) 'Locating Shima in Island Drumming: Amami Ōshima and its Archipelagic Drum Groups', *Shima*, 7(1), 14-38.

68. Jones, P. (2012) 'The Challenges of Implementing Millennium Development Goal Target 7D in Pacific Island Towns and Cities', *Asia-Pacific Development Journal*, 19(1), 139-162.

69. Kaika, M., & E. Swyngedouw (2000) 'Fetishizing the Modern City: The Phantasmagoria of Urban Technological Networks', *International Journal of Urban and Regional Research*, 24(1), 120-138.

70. Kakazu, H. (2012) *Island Sustainability: Challenges and Opportunities for Okinawa and other Pacific Islands in a Globalized World*. USA: Trafford.

71. Karides, M. (2013) 'Riding the Globalization Wave (1974-2004): Islandness and Strategies of Economic Development in Two Post-colonial States', *Island Studies Journal*, 8(2), 299-320.

72. Karlsson, A. (2009) 'Sub-National Island Jurisdictions as Configurations of Jurisdictional Powers and Economic Capacity: Nordic Experiences from Åland, Faroes and Greenland', *Island Studies Journal*, 4(2), 139-162.

73. Kelman, I., *et al.* (2012) 'Indigenous Knowledge and Disaster Risk Reduction',*Geography*, 97(1), 12-21.

74. Kelman, I., *et al.* (2011) 'Participatory Action Research for Dealing with Disasters on Islands', *Island Studies Journal*, 6(1), 59-86.

75. Kelman, I., & J.J. West (2009) 'Climate Change and Small Island Developing States: A Critical Review', *Ecological and Environmental Anthropology*, 5(1), 1-16.

76. Kong, L., & B.S. Yeoh (2003) *The Politics of Landscapes in Singapore: Constructions of 'Nation'*. Syracuse, NY: Syracuse University Press.

77. Loyd, J.M., & A. Mountz (2014) 'Managing Migration, Scaling Sovereignty on Islands', *Island Studies Journal*, 9(1), 23-42.

78. Lowenthal, D. (2007) 'Islands, Lovers, and Others', *Geographical Review*, 97(2), 202-229.

79. Maceda, E.A., *et al.* (2009) 'Experimental Use of Participatory 3-Dimensional Models in Island Community-Based Disaster Risk Management', *Shima*, 3(1), 72-84.

80. Mayaram, S. (ed.) (2013) *The Other Global City*. New Dehli: Yoda.

81. McCall, G. (1994) 'Nissology: A Proposal for Consideration', *Journal of The Pacific Society*, 17(2-3), 1-14.

82. McCusker, M., & A. Soares (Eds.). (2011). *Islanded Identities: Constructions of Postcolonial Cultural Insularity*. Amsterdam & New York: Rodopi.

83. McElroy, J.L., *et al.* (2014). 'A Note on the Significance of Geographic Location in Island Studies', *Island Studies Journal*, 9(2), 363-366.

84. McIntyre, W.D. (2012) 'The Partition of the Gilbert and Ellice Islands', *Island Studies Journal*, 7(1), 135-146.

85. Mertz, O., *et al.* (2010) 'Sustainable Land Use in Tikopia: Food Production and Consumption in an Isolated Agricultural System', *Singapore Journal of Tropical Geography*, 31(1), 10-26.

86. Mountz, A. (2014) 'Political Geography II: Islands and Archipelagos', *Progress in Human Geography*, forthcoming.

87. Neumeyer, F. (1990) 'OMA's Berlin: The Polemic Island in the City', F. Rogier (trans.), *Assemblage*, 11, 37-53.

88. Pasaogullari, N., & N. Doratli (2004) 'Measuring Accessibility and Utilization of Public Spaces in Famagusta', *Cities*, 21(3), 225-232.

89. Phua, V. (2014) 'Gazing at Haw Par villa: Cultural Tourism in Singapore',*Shima*, 8(2), 73-88.

90. Picornell, M. (2014) 'Insular Identity and Contemporary Urban Contexts: Representations of the Local in the Construction of an Image of Palma (Majorca, Balearic Islands)', *Island Studies Journal*, 9(2), 223-238.

91. Pigou-Dennis, E., & A. Grydehøj (2014) 'Accidental and Ideal Island Cities: Islanding Processes and Urban Design in Belize City and the Urban Archipelagos of Europe', *Island Studies Journal*, 9(2), 259-276.

92. Pierre, J. (2011) *The Politics of Urban Governance*. Houndmills & New York: Palgrave Macmillan.

93. Pugh, J. (2013a) 'Island Movements: Thinking with the Archipelago', *Island Studies Journal*, 8(1), 9-24.

94. Pugh, J. (2013b) 'Speaking without Voice: Participatory Planning, Acknowledgment, and Latent Subjectivity in Barbados', *Annals of the Association of American Geographers*, 103(5), 1266-1281.

95. Pugh, J., & R.B. Potter (2003) *Participatory Planning in the Caribbean: Lessons from Practice*. Aldershot: Ashgate.

96. Pugh, J. (2005) 'The Disciplinary Effects of Communicative Planning in Soufriere, St. Lucia: Governmentality, Hegemony and Space-Time-Politics',*Transactions of the Institute of British Geographers*, 30(3), 307-321.

97. Reis, A., & P. Hayward (2013) 'Pronounced Particularity: A Comparison of Governance Structures on Lord Howe Island and Fernando de Noronha', *Island Studies Journal*, 8(2), 285-298.

98. Ronström, O. (2012) 'Finding Their Place: Islands as Locus and Focus', *Cultural Geographies*, 20(2), 153-165.

99. Ronström, O. (2008) 'A Different Land: Heritage Production in the Island of Gotland', *Shima*, 2(2), 1-18.

100. Royle, S.A. (2010) 'Postcolonial Culture on Dependent Islands', *Space and Culture*, 13(2), 203-215.

101. Royle, S.A. (2009) 'Tourism Changes on a Mediterranean Island: Experiences from Mallorca', *Island Studies Journal*, 4(2), 225-240.

102. Sassen, S. (2013) 'When Territory Deborders Territoriality', *Territory, Politics, Governance*, 1(1), 21-45.

103. Sassen, S. (2006) *Territory, Authority, Rights: From Medieval to Global Assemblages*, Princeton, NJ: Princeton University Press.

104. Schrijver, L. (2006) 'The Archipelago City: Piecing together Collectivities',*OASE*, 71, 18-36.

105. Sletto, B. (2002) 'Boundary Making and Regional Identities in a Globalized Environment: Rebordering the Nariva Swamp Trinidad', *Environment and Planning D*, *20*(2), 183-208.

106. Soja, E. (2000) *Postmetropolis: Critical Studies of Cities and Regions.* Oxford: Blackwell.

107. Stratford, E. (2013) 'Guest Editorial Introduction: The Idea of the Archipelago: Contemplating Island Relations', *Island Studies Journal*, 8(1), 3-8.

108. Stratford, E., *et al.* (2011) 'Envisioning the Archipelago', *Island Studies Journal*, 6(2), 113-130.

109. Stratford, E., *et al.* (2013) 'Tuvalu, Sovereignty and Climate Change: Considering Fenua, the Archipelago and Emigration', *Island Studies Journal*, 8(1), 67-83.

110. Swaminathan, R. (2015) 'Ports and Digital Ports: The Narrative Construction änd Social Imaginaries of the Island City of Mumbai', *Urban Island Studies*, 1(1), forthcoming.

111. Swaminathan, R. (2014) 'The Epistemology of a Sea View: Mindscapes of Space, Power and Value in Mumbai', *Island Studies Journal*, 9(2), 277-299.

112. Tabucanon, G.M.P. (2014) 'Continuity and Change: Identity and Rights Protection among Later Generation Banabans', *Shima*, 9(2), 55-72.

113. Tonkiss, F. (2013) *Cities by Design: The Social Life of Urban Form*. Cambridge: Polity.

114. Tsai, H. M., & Chiang B.W. (2014) 'Enclosing Resources on the Islands of Kinmen and Xiamen: From War Blockade to Financializing Natural Heritage', *Journal of Marine and Island Cultures*, 3(2), 69-77.

115. Tutt, P.A. (2014) 'Defining the Island City: Ancient Right versus Modern Metropolis, as Considered at Peel, Isle of Man', *Island Studies Journal*, 9(2), 196-204.

116. Wachsmuth, D. (2014) 'City as Ideology: Reconciling the Explosion of the City Form with the Tenacity of the City Concept, *Environment and Planning D: Society and Space*, 31(1), 75-90.

117. Uitermark, J. (2015) 'The Particularities of Territory', *Territory, Politics, Governance*, 3(1), 1-6.

118. Ungers, O.M., & R. Koolhaas (2013 {1977]). F. Hertweck & S. Marot (eds)*The City in the City – Berlin: The Green Archipelago*. Zürich: Lars Müller.

7

# THE EUROPEAN REGIONAL DEVELOPMENT FUND AND ISLAND REGIONS: AN EVALUATION OF THE 2000-06 AND 2007-13 PROGRAMS


**Harvey W. Armstrong[1], Benito Giordano[2], Thanasis Kizos[3], Calum Macleod[4], Lise Smed Olsen[5], Ioannis Spilanis[6]**

[1]University of Sheffield, United Kingdom

[2]Aide à la Décision Économique Louvain-la-Neuve, Belgium

[3]Department of Geography, University of the Aegean, Mytilini, Greece 4

[5]Sustainable Development Consultant and Researcher Bridge of Allan, United Kingdom

[6]Nordregio, Stockholm, Sweden


## ABSTRACT


This paper presents results from a regional policy evaluation study conducted for the European Commission. The study examined the impact of the European Regional Development Fund and Cohesion Fund on EU regions with 'specific geographical characteristics', namely islands, mountain regions and sparsely populated areas. These types of regions have been attracting increasing EU regional policy attention and their economic development is considered important in helping the EU to attain its important 'territorial cohesion' objective. The focus of this paper is on the island regions. Evaluation of island regions in their own right has not been undertaken before by the EU. The study focuses on the 2000-06 and (still on-going) 2007-13 EU regional policy programs. The paper presents the methodology adopted by the study before turning to the main findings concerning the types of policy initiatives adopted in the island regions, and the appropriateness of the policies used for the economic situation faced by the islands. The islands encompassed by the study are all normal sub-national regions of EU member states. Islands with an unusual degree of administrative autonomy (e.g. the Outermost Regions) were excluded.

## KEYWORDS

European Union; evaluation; island regions; policy initiatives; regional policy

## 1. INTRODUCTION

Interest on how European Union (EU) regional policy can be best brought to bear in islands has been growing in recent years. This interest is likely to continue to grow because the performance of regions with specific geographical features is considered to be an important element in achieving the EU's key long term regional policy objective of territorial cohesion (European Commission, 2006).

While the focus on islands is relatively new, the EU has a long history of devising distinctive policies for regions with specific geographical features. Special assistance for the Outermost Regions (ORs – the Azores Islands and Madeira of Portugal, the Canary Islands of Spain, and the French overseas departments of Guadeloupe, Martinique, Réunion and recently Mayotte – all islands – and Guyana) has been in place for decades (European Commission, 2012a). The specific 'geographical feature' which triggered the help for the ORs is remoteness from the EU market. The other type of 'specific geographical feature' region which has attracted more recent special EU intervention has been sparsely populated areas (SPAs). The EU accession of Austria, Finland and Sweden in 1995 resulted in SPAs being accorded special status within EU regional policy (European Commission, 2006), and they have enjoyed this ever since.

This paper presents results from a recent study undertaken for the European Commission (Aide à la Décision Économique – ADE, 2012). The study was an evaluation of the 2000-06 and 2007-13 European Regional Development Fund (ERDF) programs as they have been applied in regions with 'specific geographical features': namely islands, mountain regions and sparsely populated areas. The paper focuses on the island regions. It begins with a brief review of the methodology adopted for the study. The key results for the island regions selected for evaluation are then presented. The Conclusion draws out the main findings and summarizes key similarities and differences between the island regions and the other mountain and SPA regions analysed.

## 2. EVALUATION METHODOLOGY AND SELECTION OF THE CASE STUDY REGIONS

The evaluation adopted a case study methodology. Case study evaluation is one of a suite of evaluation methods recommended by the European Commission (European Commission, 2012b), and is principally a qualitative evaluation methodology. There are two key advantages to using a case study methodology for island regions:

a) Existing quantitative data sets (e.g. Eurostat's harmonized statistics) are extremely deficient when it comes to islands. This makes modern quantitative counterfactual impact evaluation almost impossible without very expensive surveys (Martini, 2009). Just how hampered researchers are in seeking to undertake cross-island statistical analysis is vividly shown by those studies which have been attempted. For example, a Planistat Europe (2006) study was able to analyse only 19 of the 286 EU islands with over 50 inhabitants. Moncada et al. (2010) were able to analyses only 28 of 5,116 islands in 28 European countries. An ESPON (2011) study was able to analyse only 31 of 362 inhabited EU27 islands with populations over 50 persons. These 31 islands comprise those EU statistical regions "for which some common basic data are available" (ESPON, 2011, p.5). The important study by Monfort (2009) was able to analyse only 56 islands. These comprise the set of EU statistical regions made up wholly of islands. Extracting data for individual islands from large harmonized data sets such as Eurostat is extremely difficult for three reasons: (i) many of the islands are so small that researchers must use data from only the most highly geographically disaggregated parts of the data-base. At these levels, the data is fragmentary and often simply non-existent for key variables; (ii) at broader levels of geographical aggregation the islands are frequently grouped in with parts of the mainland littoral, making it impossible to extract data for the island part of the regions; and (iii) at higher levels of geographical aggregation data are often only available for groups of islands and not individual islands.

b) The case study approach offers a better chance of overcoming an important problem faced by quantitative evaluations of ERDF programs. These programs are delivered through multi-stakeholder partnerships, involving different national, regional and local organizations. Case studies are arguably better than econometric methods at disentangling the effects of different partners and funding streams. This is because the case study technique "draws on a multitude of perspectives..... or derives from multiple accounts of different actors in the setting ... Case studies are information rich ... The defining feature of the case study is that it is information rich" (European Commission, 2012b, Evalsed Sourcebook: Methods and Techniques: Case Studies Section). This advantage of the case study approach is particularly valuable for those types of policies which involve many different actors and partners, such as innovation policies (Technopolis Group, 2012) or local and community level policies (Instituto per la Ricerca Sociale, 2011). Policy interventions in island economies also involve a wide array of economic and social stakeholders and hence are also prime candidates for a case study methodology. On the downside, a problem faced by case study methods is that, because they are so costly (in time and money) to undertake, it is difficult to generalize from just a few cases to a full set of regions. There has been a lively debate concerning how generalizable case study evaluations are (Flyvbjerg, 2006; Koenig,

2009). This inherent weakness means that great care must be taken in the selection of the case study regions.

In this study, a two-stage procedure was adopted. At the first stage of the study, a set of 15 NUTS2 regions with 'specific geographical features' were chosen. These were subject to a broad brush analysis, using statistical information and a review of previous evaluations and studies of the regions. This broad overview of 15 regions formed the basis of a second stage at which six regions (at NUTS3 or lower) were selected for in-depth analysis. The numbers of case study regions at each stage were pre-set by the European Commission's terms of reference and were determined by the available financial resources for the evaluation study. The NUTS regions are those based on the EU's standard nomenclature for regions for statistical data collection. The hierarchy of regions runs from NUTS0 (the Member States) down to NUTS5 at a very local level. An unfortunate feature of NUTS2 regions is that most tend to group islands with parts of the mainland littoral. This makes the NUTS2 regions peculiarly unhelpful concepts for studying islands.

The criteria for selection of the 15 regions for the initial, 'broad brush' overview were designed to produce as representative a group of regions as possible across the three categories of region (i.e. island regions, mountain regions and sparsely populated regions). Additional criteria for selection were: (a) the list should include both Objective 1 and Objective 2 regions, (b) at this first stage should be NUTS2 regions to maximize data availability, (c) the list should include at least one region from each member state specified by the European Commission in its tender specifications, (d) should preferably comprise NUTS2 regions encompassing several NUTS3 regions within them (to maximize choice at the second stage), and (e) should take into account likely data availability at NUTS3 level (to assist stage 2 in-depth analysis). In practice, the precise selection of the 15 initially selected regions and the six in-depth case study regions was the result of discussions with the European Commission.

Of the 15 initially selected regions, six were wholly or partially island regions (Balearic Islands, Corse, Sicilia, Voreio Aigaio, Hovedstaden – including Bornholm, and the Scottish Highlands & Islands – H&I). As noted above, from these, six regions were selected for indepth analysis (at NUTS3 level or lower). These comprised Cuenca (within Castilla la Mancha in Spain); Norrbotten (within Övre Norrland in Sweden); Lesvos (within Voreio Aigaio in Greece); Bornholm (within Hovedstaden in Denmark); Ardèche (within Rhône-Alpes in France);

and the Western Isles (within the H&I region of the UK). Three of these are island regions: Bornholm, Lesvos and the Western Isles

There has been a long debate on how mountain regions, islands and SPAs should be defined (e.g. Gløersen et al., 2006; Planistat Europe, 2006; Monfort, 2009). The EU has developed its own particular definitions (Monfort, 2009). Mountain regions are defined using a combination of measures of topography (altitude and slopes), the dominant climatic condition (i.e. contrasts in temperature) and land use (i.e. forests, moorland). SPAs are defined using simple population density measures. Islands are defined using a combination of minimum population size and area, minimum distance from the mainland, the absence of a fixed link to the mainland and must not contain an EU member state capital, except Malta and Cyprus (Monfort, 2009).

The study drew upon a combination of research techniques. These included some (limited) cross-EU analysis of regional statistics, an extensive literature review, a review of policy and previous evaluation documents, and an intensive program of interviews and focus groups in each of the six in-depth case study regions (conducted in the second half of 2011). Interviewees and focus group members were selected from amongst the key stakeholders engaged in the programs using two approaches: (a) key managers of the programs themselves were pre-selected and interviewed, followed by (b) a 'snowball' approach to identify and interview other key stakeholders. By 'key' is meant those engaged in the main policy interventions, together with those best informed of how the programs have evolved since 2000 (an important criterion give staff turnover across the two program periods). The full lists of interviewees can be found in annexes to the case study reports and included private, public and voluntary sector stakeholders. Presentation of the results was done using a narrative format, as is appropriate with the case study evaluation methodology (Constantino & Greene, 2003; Yin, 1981).

Finally, it should be noted that the various regions with 'specific geographical features' do not comprise a mutually exclusive classification. Indeed, many EU islands are also mountainous and/or sparsely populated. Figure 1 shows the three in-depth island case study regions which are the focus of this paper. The Western Isles are both an island region and also a SPA. Part of the region is also mountainous. Lesvos is both an island and mountainous. Only Bornholm is solely an island region.

A serious problem facing the research from the start was the absence of a single consensus theory of economic growth for regions. Regional policy interventions such as those of the ERDF typically seek to trigger faster economic growth. They are essentially 'supply-side' policies which rely principally for their success not on the fiscal expenditure stimulus of new spending, but rather on encouraging growth by removing impediments or creating new opportunities. This in turn relies upon there being a good understanding of the underlying determinants of economic growth. However, at the present time, the determinants of economic growth for islands, as with other types of sub-national regions, remain only partially understood and a number of different economic theories have been proposed. These range from traditional neoclassical growth theory through export-led models to newer theories such as endogenous growth theory and new economic geography models (see Armstrong & Taylor, 2000 for a review). In the absence of a single growth theory, the study adopted the more pragmatic approach of a SWOT (strengths, weaknesses, opportunities and threats) framework. This has two advantages. Firstly, it allowed a holistic approach to be adopted, encompassing as full a set as possible of growth determinants. Secondly, it allowed the study to draw upon ERDF strategy program documents (called Single Programming Documents – SPDs, or alternatively Operational Programs - OPs) which are routinely based on detailed SWOT analyses for each region.

*Location of Voreio Aigaio*
*(Source: Eurostat)*

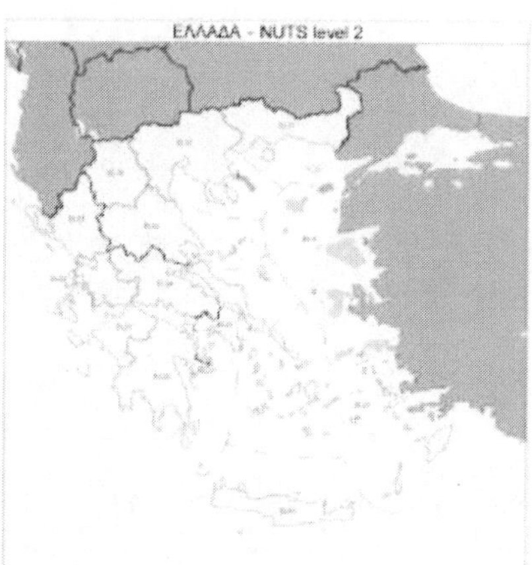


*Location of island of Lesvos*
*(Source: Eurostat)*

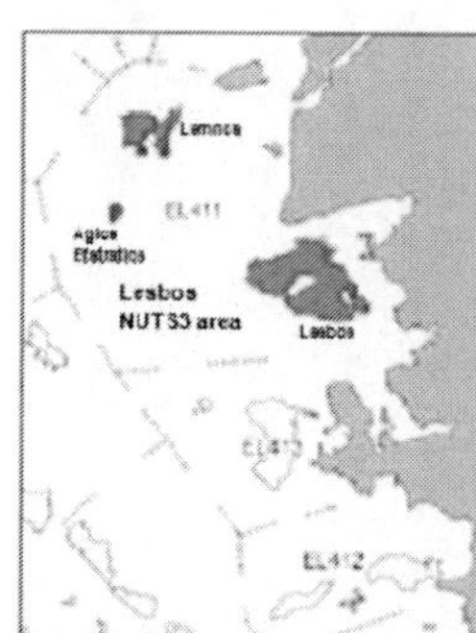

*Location of Hovedstaden*
*(the Copenhagen capital region,*
*to which Bornholm belongs adminstratively)*
*(Source: Eurostat)*

*Location of the island of Bornholm*
*(Source: Eurostat)*

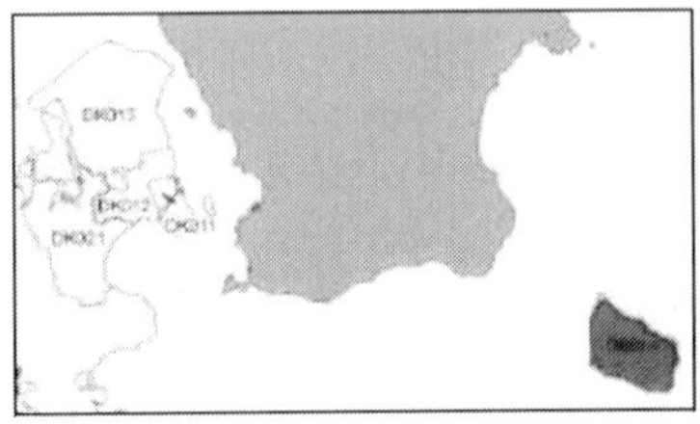

*Location of the Highlands and Islands*
*(Source: Eurostat)*

*Location of Western Isles of Scotland*
*(Source: Eurostat)*

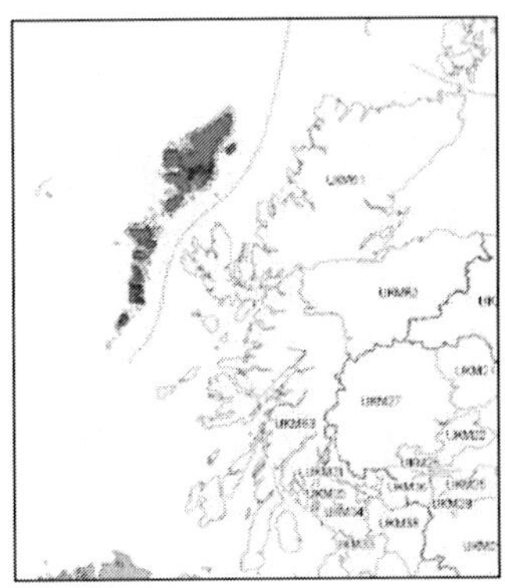

**Figure 1.**Location of the three case study islands and their NUTS2 regions

Detailed SWOT checklists were developed for island regions, mountain regions and SPAs (see ADE, 2012, Second Intermediate Report, Annexes 1-3). The SWOT lists formed the basis of the semi-structured interviews, focus groups and other fieldwork.

# 3. RESULTS FOR ISLAND CASE STUDY REGIONS

The results for the three island case study regions have been grouped under three main headings:

**1. Key geographical characteristics.**

An important conclusion of the study was that, from the perspective of the principal policymakers and stakeholders, geographical specificities do indeed matter. This need not necessarily have been the case as most EU islands lie close to the European landmass and rapid improvements in transport and telecommunications have undoubtedly eroded the economic challenge posed by islandness (we follow Baldacchino, 2007a, in using 'islandness' rather than 'insularity' as the latter term has pejorative connotations). The case studies do, however, reveal that the influence of geography is more complex than might be thought, for four reasons:

a) In no case was islandness found to be the sole source of economic challenges. In all three island regions, it was a combination of geographical characteristics which mattered;

b) Remoteness (from EU markets) and the settlement pattern (particularly the widespread characteristic of 'small scattered settlements') were revealed as being of particular importance. Mountainous terrain does not seem to pose such a serious challenge and was frequently seen as a positive attribute;

c) For some of the islands, the problem of 'small scattered communities' was made worse by the geographical configuration of the landscape;

d) Each of the three case study regions exhibited their own unique 'accident of geography' which generated further distinctive challenges.

The case study which best exemplifies all four of these subtle interactions of geographical characteristics was the Western Isles. The Western Isles lie off the north western seaboard of Scotland, making them some of the remotest of all EU islands. In a study of the accessibility of each of NUTS3 regions, Copus (1999) derived a Western Isles accessibility index of 89.53 (on a scale of 0-100), a very high value.

More recent accessibility analysis places the Western Isles firmly in the category of 'predominantly rural, remote regions' (Dijkstra & Poelman, 2008, 2011).

To the challenges of islandness and remoteness in the Western Isles must be added the phenomenon of sparse population. The Western Isles population density in 2007 was only 8.6 persons per km2 ; the equivalent figure for population density across the wider EU in 2007 was 113 persons per km2 . Typically, however, the main challenge of sparse population was not low population density itself, but rather a pattern dominated by small scattered communities. Economic activities struggle to survive in small, widely separated communities and public services are costly to provide. The traditional settlement pattern of the Western Isles was one of tiny rural 'townships'. The challenge posed by the pre-existing pattern of scattered settlements has been exacerbated by recent demographic trends. In a trend typical of many EU islands, increasing 'primacy' in the urban hierarchy has occurred. The islands' capital, Stornoway, has become increasingly dominant and now contains 30% of the total population of the region. Increasing primacy has occurred against a backdrop of falling total population over time (a fall of 43% between 1901 and 2001 and no less than 8.5% between 1995 and 2005). As a result, individual settlement sizes away from Stornoway have fallen.

The inherent tendency towards small scattered settlements is made worse by an unusual geographical configuration of the landscape, in three ways:

a) The Western Isles are made up of an archipelago of 100 or so islands, resulting in issues of 'islands off islands', or 'double insularity' (Bardolet & Sheldon, 2008). Depopulation of the smaller islands has led to 13 currently inhabited islands, and the construction since the 1950s of a 'Spinal Route' of bridges and causeways has reduced the archipelago to only three groups of inter-connected islands. Nevertheless, 'double insularity' and the fragmentation of the settlement pattern remains a serious issue;

b) The landscape is one of a complex pattern of sea lochs and peninsulas, further fragmenting the pattern of small scattered settlements;

c) The archipelago's islands straggle in a long north-east to south-west chain stretching 100 km (giving rise to one of its earliest names: the 'Long Isle'), again making access within the islands difficult.

There is also the Western Isles' own unique 'accident of geography'. The islands are served by two main ferry routes. Unfortunately, the mainland coast facing the islands is itself both sparsely populated and a considerable distance from the H&I capital of Inverness (on the east coast of Scotland) or other major Scottish cities. Long overland transport must accompany each ferry journey. This situation is ameliorated somewhat by improvements in air transport links over the last twenty years or so in the Scottish islands. The Western Isles have three airports (at Barra in the south, Benbecula in the centre of the chain and at Stornoway in the north, the islands' main airport). There are scheduled inter-island flights as well as direct links to Inverness and Glasgow. While improvements in air transport have undoubtedly greatly reduced the isolation faced by islanders, as is the case for many islands around the world, air transport can only ever partially offset sea transport challenges. Air transport is expensive and is only suitable for certain types of freight and passenger journeys. The Western Isles' inhabitants do, however, benefit from cheaper air fares subsidized by the Scottish government (40% reductions for registered local inhabitants), but these subsidies do not apply to visitors.

The challenges posed by a combination of geographical characteristics can often be subjective as well as objective. For example, the perception of a challenge such as remoteness can be as important as its (declining) reality: "There is also the issue of the perception of remoteness, both within the region and by outsiders, even though transport access has greatly improved over the years" (Interviewee). This survival of a perception of islandness and remoteness despite major transport improvements has been identified elsewhere in island communities (e.g. Baldacchino, 2007b).

Lesvos is one of nine main islands comprising the Greek NUTS2 region of Voreio Aigaio. Although Voreio Aigaio is not the same kind of free-standing archipelago found in the Western Isles, Lesvos does face major 'double insularity' problems as well as a combination of islandness, remoteness and settlement pattern challenges. The island lies at the extreme eastern periphery of both Greece and the EU. Even the most direct ferry route to Athens is a long one (12 hours on the main route), and Lesvos has a Copus (1999) accessibility index of 95.32, more remote even than that of the Western Isles. As with the Western Isles, Lesvos has benefited from improvements in air transport links. Mytilíni has a modern international airport and there are direct charter flights from no fewer than eight northern European countries as well as regular scheduled flights to Athens and Thessaloniki. While superficially Lesvos does not seem to have a sparse population challenge, having a population density of 189.5 persons per km2 , it does face an issue of small scattered settlements arising from (a) the configuration of the landscape (it is a mountainous island with a coastline of peninsulas and inlets), and (b) increasing urban primacy around Mytilíni, the administrative capital of the island. The latter has once more occurred against a backdrop of depopulation (a 35% loss of population between 1951 and 1981, a small recovery of 3.4% between 1981 and 2001, but another drop of 4% between 2001 and 2011). The more rural settlements have been in decline, at least up until the 2008 financial crisis. There is some anecdotal evidence that return migration to smaller remote settlements may have begun to pick up in the aftermath of Greece's savage austerity program, but it is still too early to know how large or how permanent this might prove to be in practice.

Finally, there is Lesvos' own unique 'accident of geography'. The Turkish mainland is extraordinarily close (a mere 5.5 km at its nearest point). While the Turkish mainland opposite Lesvos is not sparsely populated, it might as well have been. Turkey is not an EU Member State and for historical reasons economic links between Lesvos and Turkey have been strictly limited, even though in recent years some Turkish tourists have been able to visit the island.

Bornholm does not face the same number of different geographical challenges as the Western Isles or Lesvos. It is not part of an archipelago. Nor is it mountainous or sparsely populated, having a population density of 71.2 persons per km2 . The island is also a compact one of 589 km2 , with good internal transport links from smaller communities to the capital Rønne. The configuration of the landscape is therefore not the issue it is in Lesvos or the Western Isles. On the other hand, it does face a combination of islandness and remoteness (with a high Copus index of 83.29). Moreover, settlement sizes in the rural areas have been affected by depopulation. Bornholm's population peaked in 1965 (at 48,620 persons), but has fallen by 14% since then.

Finally, there is Bornholm's own unique 'accident of geography'. Bornholm is a free-standing small island lying in the southern part of the Baltic Sea. The island is much closer to Sweden (37 km), and even Poland (90 km) and Germany (88 km) than to the Danish capital city region of Hovestaden of which it is a part (145 km). As a result, the island's main ferry route is with another member state (i.e. through Ystad in southern Sweden), and as a result of this quirk of geography it is actually quicker to travel from Bornholm to Copenhagen via another Member State than directly! Once again, as with the Western Isles and Lesvos, improvements in air transport have helped to overcome the 'tyranny of distance' imposed by sea transport. Bornholm has regular scheduled flights directly to Copenhagen and there is a program of summer flights to and from Berlin. However, once again it should be noted that air transport is both an expensive option and also only a suitable alternative to the ferry services for certain types of freight and passengers.

## 2. Niche market specialization

Virtually all small island economies are fated to rely on niche markets for their economic prosperity. This is a double-edged sword. Specialization can lead to great prosperity. On the other hand, it also renders the economies vulnerable to sudden shifts beyond the island's control.

While all the case study regions show clear niche specialization, they also reveal three other features:

a) Existing niche sectors are insufficient to maintain high income levels and full employment. This is, of course, why all three are designated EU regional policy regions, but it is also true that a majority of EU island regions also seem to underperform (Monfort, 2009);

b) Islands do not enjoy the luxury of being able to stand still. In all three regions, it has proved necessary to seek new niches as traditional niches have declined. In the best cases, the islands have sought to build from their strengths and move forward into higher value added niches within the existing sectors. In other cases, this has not proved possible and economic performance has suffered accordingly;

c) Diversifying away from traditional niche sectors into radically new types of business has proved to be extremely difficult. Despite major efforts to do this, progress has been relatively slow in all three regions.

It is interesting to note that these three findings of the study are strongly supported by the wider research literature dealing with remoteness, insularity and vulnerability (e.g. Briguglio et al., 2006; Baldacchino & Milne, 2000; Eurisles, 2002).

The three case study regions exhibit fascinating differences in how they have sought to develop new niche markets. The Western Isles has a distinctive set of niche sectors. The islands lack oil & gas and timber resources (unlike other parts of the H&I region), but do have two very valuable resource-based niches: fishing and fish farming. The former is dominated by high value shellfish (mostly live prawns for shipment to Europe - some 3,200 tonnes in 2009 worth £10.2m - US$15.9m at current exchange rates). The latter is dominated by Atlantic salmon fish farming in the islands' sea lochs, generating some 350 local jobs directly and another 200 in processing, marketing and distribution. There are three other main niche sectors:

a) Tourism, principally activity and 'cold-water' tourism (Baldacchino, 2006), worth £50m (US$78m) in 2007 and generating 1,250 jobs (Snedden Economics, 2007).

b) Agricultural products, with exports dominated by 'store' sheep and cattle (i.e. animals reared for fattening elsewhere on the mainland), worth £1.2m (US$1.9m) in export earnings in 2003.

c) Textile, craft and creative industries, based on the islands' distinctive culture. Most notable among these is the world-famous Harris Tweed textile industry (worth £3.9m, or US$6.1m, in 2009).

The Western Isles are fortunate in having such a wide array of niche sectors. However, all of the key sectors reveal great vulnerability. Fishing was traditionally dominated by white fish catches. The introduction of strict UK and EU quotas in the 1990s put paid to this valuable niche. The islands then switched to shellfish and fish farming. However, even this has proved volatile. Fish farming expanded rapidly in the 1990s, followed by a large decline in prices leading to a period of employment losses. Tourism has proved to be more stable, but shifts in UK tourism away from traditional summer holidays within the UK has meant that other forms of tourism have had to be fostered. Finally, the Harris Tweed industry has also proved vulnerable to external fashion shifts. In the 1990s and early 2000s, demand and production fell sharply before partially reviving after consolidation and major efforts to rebrand the niche as a fashion product (e.g. in handbags).

As with many small islands, declining job opportunities in the main niche export earning sectors have been compensated in part by expanding public sector employment. In 2008 employment in public administration, education and health amounted to 42.6% of all jobs in the Western Isles, compared to only 32.8% in the H&I as a whole (Highlands & Islands Enterprise, 2009).

While the Western Isles have been able to adapt to changes in their traditional niche sectors, it has proved much harder to develop radically new niches. For example, development of renewable energy for export has proved difficult despite EU support, and the craft and cultural niches (e.g. Gaelic media enterprises) remain very small.

Lesvos and Bornholm give us two very different examples of island adaptation. The Lesvos economy relies most heavily on only two sectors – agriculture (or rather agri-business through food processing) and tourism. Fishing is a small and declining sector. Within the two key export sectors, great sub-sector specialization exists. Agri-food products are dominated by olive oil (with olive trees covering some 40% of the island), soap production, ouzo spirit production, cheese and specialist products such as gum mastic. Unlike the Western Isles, tourism is mainly of the traditional mass 'sun and sand' category. The agri-food business explains the relatively high proportion of manufacturing in GDP (5.6% in 2007 compared to 4.8% in the wider Voreio Aigaio region, though still small compared to the Greece average of 10.4%), and also the higher than average contribution of agriculture to GDP (5.8% in 2007 compared to 3.8% for Greece as a whole). Tourism has been the most stable of the niche sectors, but not very dynamic. Agri-food products have faced greater challenges. For example, olive oil and soap production sectors began experiencing export competitiveness pressures from the 1950s onwards, and high labour costs in Greece have only been partly offset by inflows of eastern European migrants. In summary, Lesvos has found itself locked into a small set of traditional niche sectors and has struggled to adapt to external changes.

In the case of Bornholm, the economy rests most heavily on summer tourism, food products (particularly dairy products), some limited manufacturing (including metal working and crafts and pottery production) and fishing. The four key export sectors generated directly some 4,600 jobs in 2008 (22.7% of the total), although many more jobs rely indirectly on export income. Between 1996 and 2008, employment in the traditional agriculture, food processing and fishing sectors fell. Production levels in agriculture, manufacturing and tourism niches have fortunately held up. The best example of niche volatility has been fishing. This was a major employment and export sector until the 1980s, when the introduction of quotas in response to declining fish stocks greatly affected the industry. As shall be shown later, of all three island case studies, Bornholm has proved to be the most radical in seeking to develop new niches, particularly in 'green tourism', new industrial clusters and renewable energy. This has not, however, been easy.

### 3. The demographic challenge

The most striking feature revealed by the study for all three island case studies is the scale of the demographic challenge of depopulation. The settlement pattern implications of falling population have already been discussed. There are, however, other more important impacts. Depopulation undermines island industries and puts public service provision under severe pressure. More serious still is the loss of younger, more economically active people. In Bornholm, population decline since 1965 has been greatest among 20-29 year olds. This in turn has triggered a fall in birth rates too, leading forecasters to predict a further population fall to 39,470 inhabitants by 2021. This vicious circle, with an ageing population that is becoming ever more elderly, is occurring in all three case study regions.

Lesvos has had almost continuous population decline since 1951. A temporary population recovery of 3.4% over 1981-2001 was the result of return migration and in-migration flows from Eastern Europe rather than natural increase. Deaths have exceeded births on Lesvos for the last 50 years. By 2001, Lesvos had an 'ageing index' (i.e. people over 65 divided by people under 15 years of age) of 145, compared to a Greek national average of 110.

The Western Isles have faced even bigger problems. Population has fallen continuously since 1901, falling 43% between 1901 and 2001. The period 2001-2009 saw a further fall of 1%. Between 2001 and 2008, the percentage aged 65 and over rose from 20% to 21%, while those under 44 fell from 53% to 50%: "the demographics for the Western Isles still look horrendous ... The issues are just as stark as ever" (Interviewee). Some in-migration for 'lifestyle' reasons has occurred in the Western Isles (as in the other two regions), but is nowhere near enough to offset the loss of young people (Hall Aitken, 2007).

### 4. The ERDF programs in 2000-06 and 2007-13.

ERDF policy initiatives were classified in the study into three broad groups:

a) Reactive policies – those defensive in nature and designed to mitigate or offset particular geographical weaknesses or threats (e.g. transport links);

b) Proactive policies – designed to exploit island strengths and opportunities (e.g. craft enterprises exploiting island culture);

c) Sustainability policies – those seeking to exploit islands as 'laboratories' for developing more environmentally sustainable economic development. None of the three case studies have sought to go as far as Samsø in Denmark which has a fully carbon neutral strategy (Samsø Energy Academy, 2007); but all three case studies have major sustainability components.

This classification has the advantage of pin-pointing the extent to which an island has managed to move on from traditional views of islandness as a set of problems to more forward looking policies seeking to build on island strengths.

The 2000-06 and 2007-13 programs differ greatly in scale between the three regions. Lesvos is part of the Voreio Aigaio NUTS2 region and was eligible in both 2000-06 and 2007-13 for help as part of an Objective 1 ('Convergence') region. Objective 1 regions are much more generously funded than Objective 2 ('Regional Competitiveness and Employment') regions. Uniquely amongst the three case studies, it was also eligible for Cohesion Fund (CF) assistance, Greece having a GDP per capita under 90% of the EU average. The Western Isles are part of the H&I region of Scotland and in both 2000-06 and 2007-13 were eligible for 'transition' funding (having been an Objective 1 region in 1994-99). Bornholm was part of Denmark's sole Objective 2 program in both periods, and hence had the least generous ERDF funding of all three case studies.

Lesvos, and the wider Voreio Aigaio region of which it is part, was not just better funded in both 2000-06 and 2007-13, but also experienced an increasing EU budget. ERDF/CF allocations for Voreio Aigaio rose from €358.5m to €373m between 2000-06 and 2007-13. In Denmark, overall EU allocations for the Objective 2 program were €137m in 2000-06, of which Bornholm received €8.5m. For the 2007-13 period, Denmark's allocation for the Regional Competitiveness and Employment objective rose to €254m, although Bornholm's share of this fell to €7.5m. By contrast, ERDF allocations for the H&I region fell from €190m in 2000-06 to €122m in 2007-13.

An important feature of ERDF programs in the three regions concerns their governance arrangements. In all three cases, the islands had to function within programs designed for a wider eligible region. This caused major difficulties in tailoring the wider economic development strategy to the particular needs of the islands, and especially so for Lesvos. This island was part of the Voreio Aigaio administrative region in both periods, but the 2000-06 Objective 1 program actually covered the whole of Greece. While separate broad regional programs were developed within the national Objective 1 program, they were at a level well above that of Lesvos, and the Greek Objective 1 program also had a series of sectoral programs running at national level (e.g. transport). The 2007-13 Operational Program, of which Lesvos is part, covers several different Greek NUTS2 regions, broader even than Voreio Aigaio, and again there are some national level sector initiatives.

The Western Isles has had rather more policy latitude than Lesvos, but shared the same conundrum of being part of a wider (H&I) geographical program. Moreover, H&I programs were designed with mainland areas in mind as well as islands. Bornholm has had the most freedom of all the case study regions to develop a more tailored program. This is because it has been granted the unusual status within the Danish system of a combined regional and local government, and in 2007 was granted its own 'Regional Forum' of partners to design economic development strategy. It was also given considerable freedom within the ERDF programs themselves to allocate the funding. Nevertheless, even in the Bornholm case, the Objective 2 program is essentially a national one, of which Bornholm is only a small part.

Since in all three cases the islands are part of wider ERDF program areas, a key consideration is how the islands' shares of the ERDF budget are determined. In the case of the Western Isles a deliberate decision was taken not to divide up the 'cake' into pre-determined shares, but rather to allow each sub-region's share to be determined by the quality of projects eventually put up for approval. During 2000-06, the Western Isles obtained some €13m (or 7% - US$16.1m) of the full H&I spending, equivalent to the value of €494 per capita (US$612), well in excess of the H&I average of €429 (US$532). Equivalent figures for 2007-13 are not yet available as the program is on-going. In the case of Lesvos and Bornholm, more formal procedures were adopted to guarantee their budget shares. Different islands within Voreio Aigaio are allocated shares based on population size, giving Lesvos a 52% share in both 2000- 06 and 2007-13. Bornholm was pre-allocated a share of 6.6% in 2000-06. This fell, however, to only 2.9% in 2007-13.

Turning to the nature of the ERDF interventions, commitments can be analyzed by broad 'fields of intervention'. These are shown in Table 1 below.

**Table 1.** Allocations by Field of Intervention, 2000-06.

| *Field of intervention* | *EU Obj. 1* | *Voreio Aigaio* | *Lesvos* | *EU Obj. 2* | *Denmark Obj. 2* | *Bornholm* | *H&I Region* |
|---|---|---|---|---|---|---|---|
| Agriculture | 0.1 | 0.0 | 0.0 | 0.2 | 0.0 | 0.0 | 0.0 |
| Forestry | 0.0 | 0.0 | 0.0 | 0.1 | 0.0 | 0.0 | 0.0 |
| Adaptation and development of rural areas | 0.4 | 0.0 | 0.0 | 2.5 | 4.1 | 0.3 | 0.0 |
| Fisheries | 0.1 | 1.6 | 0.3 | 0.0 | 0.0 | 0.0 | 0.0 |
| Assisting large businesses | 5.8 | 0.7 | 0.6 | 5.1 | 2.3 | 1.6 | 0.0 |
| Assisting SMEs and craft firms | 9.6 | 5.1 | 5.2 | 31.2 | 28.6 | 33.8 | 35.6 |
| Tourism | 2.9 | 5.4 | 4.0 | 10.2 | 30.4 | 1.6 | 4.8 |
| Research, technological development and innovation (RTDI) | 6.2 | 2.0 | 1.9 | 10.1 | 14.6 | 6.9 | 3.6 |
| Labour market policy | 0.1 | 0.0 | 0.0 | 0.1 | 0.0 | 0.0 | 0.0 |
| Social inclusion | 0.1 | 0.0 | 0.0 | 0.7 | 0.0 | 0.0 | 0.0 |
| Education and vocational training | 1.5 | 0.7 | 0.7 | 1.4 | 0.0 | 0.0 | 0.0 |
| Workforce flexibility, entrepreneurial activity, innovation and ICT | 0.0 | 0.0 | 0.0 | 0.4 | 0.0 | 0.0 | 0.0 |
| Positive labour market actions for women | 0.0 | 0.0 | 0.0 | 0.1 | 0.0 | 0.0 | 0.0 |
| Transport infrastructure | 34.0 | 26.8 | 24.1 | 7.7 | 7.4 | 14.2 | 31.9 |
| Telecommunications infrastructure | 3.2 | 16.2 | 14.8 | 2.9 | 5.2 | 33.3 | 6.7 |
| Energy infrastructure | 1.0 | 0.0 | 0.0 | 0.8 | 0.2 | 0.5 | 4.0 |
| Environmental infrastructure | 19.3 | 27.1 | 38.8 | 5.2 | 0.0 | 0.0 | 3.2 |
| Planning and rehabilitation | 10.1 | 8.6 | 6.2 | 17.0 | 5.0 | 4.2 | 0.0 |
| Social and public health infrastructure | 4.2 | 4.9 | 3.0 | 1.9 | 0.0 | 0.0 | 7.6 |
| Technical assistance | 1.4 | 0.9 | 0.4 | 2.3 | 2.0 | 3.5 | 2.5 |
| *Total* | *100.0* | *100.0* | *100.0* | *100.0* | *100.0* | *100.0* | *100.0* |
| Productive environment | 25.1 | 14.7 | 12.0 | 59.5 | 80.0 | 44.3 | 44.1 |
| Human resources | 1.8 | 0.7 | 0.7 | 2.7 | 0.0 | 0.0 | 0.0 |
| Basic infrastructure | 71.7 | 83.6 | 86.9 | 35.6 | 17.9 | 52.2 | 53.4 |
| Technical assistance | 1.4 | 0.9 | 0.4 | 2.3 | 2.0 | 3.5 | 2.5 |

Notes: 1. 'Human resources' (i.e. labour market policy and other initiatives in the central part of the table) is essentially an ESF responsibility and is thus not considered further here.
2. Separate data do not exist for the Western Isles.
3. Figures are percentages of the total ERDF/CF budget.

Three key features emerge from Table 1:

ERDF policies in Lesvos in 2000-06 were very different from those in Bornholm and in the Scottish H&I region. In particular, Lesvos saw much more of the budget targeted on basic infrastructure projects (86.9%) than either Bornholm (52.2%) or the H&I region (53.4%). This is partly a reflection of the fact that Lesvos is an Objective 1 region. Objective 1 regions have larger grant rates for infrastructure projects and EU regional policy encourages greater infrastructure spending there. However, Lesvos has clearly taken a deliberate decision to adopt a stronger infrastructure element than other EU Objective 1 regions (71.7%), and has an even higher infrastructure commitment than the Voreio Aigaio region of which it is part. As can be seen, in Lesvos three main types of infrastructure projects have prevailed: environmental infrastructure such as water supply (38.8%), transport infrastructure (24.1%) and telecommunications infrastructure (14.8%).

In contrast, Bornholm and the H&I region committed much more of the budget in 2000-06 to 'softer' productive environment initiatives (e.g. business advice, grants and loans to firms etc). These made up 44.3% of the budget in Bornholm and 44.1% in the H&I region. The switch from infrastructure to business support is what one would expect for all Objective 2 regions ('Growth and Competitiveness' regions). The particular focus in Bornholm and H&I on SMEs and craft firms (33.8% and 35.6% of the budget respectively) is noteworthy, being greater than the overall Objective 2 average (31.2%).

However, the two Objective 2 island regions do still have a lot of basic infrastructure spending. Bornholm put 52.2% of its 2000-06 budget into basic infrastructure while the equivalent figure for the H&I was 53.4%. The overall EU Objective 2 figure was only 35.6%. Table 1 shows that there was a big difference between Bornholm and the H&I region in the type of infrastructure assisted. Bornholm was much more focused on telecommunications (a massive 33.3%) while the H&I region favoured transport infrastructure projects.

In summary, in 2000-06 the island regions tended to put a lot more emphasis on infrastructure projects than their wider EU counterparts, and amongst the three case studies Lesvos was extremely heavily committed to infrastructure policies, while Bornholm and the H&I region had by 2000-06 moved on to somewhat more of the softer business support policies.

Turning to the 2007-13 program period, the new regional strategies were developed against a backdrop of the European Commission requiring regional policy to focus much more closely on the so-called Lisbon Agenda objectives of better business competitiveness, growth and environmental sustainability, implying more help for softer business support measures and less for basic infrastructure. In practice, this change was much more prominent among Objective 2 regions than Objective 1 areas such as Lesvos.

In order to ascertain whether the Western Isles have followed the desire to embrace the Lisbon Agenda, an analysis was undertaken of the actual projects approved by the time of the study (August 2011). Projects were classified as reactive, pro-active or sustainability ones. The exercise was conducted separately for the 17 Western Isles ERDF projects approved by August 2011 under the 2007-13 program as well as for the full set of projects approved for the complete H&I region of which the Western Isles are a part. Table 2 presents the results. This exercise was done only for the Western Isles and not for Bornholm or Lesvos.

Caution must be exercised in interpreting the results, for two reasons. Firstly, the 2007-13 program still has some distance to run and the figures in the table refer only to projects approved by August 2011. However, this is much less of a problem than might appear at first sight since by August 2011 the vast bulk of the EU funds had been committed. Secondly, Table 2 does not include figures for ERDF money spent by Highlands & Islands Enterprise (HIE), the government-funded regional development agency. HIE was block allocated funding at the start of the program for use within their already established business support programs for firms. Separate figures for Western Isles involvement in HIE projects were collected (see below). These refer for the period up to 15 June 2011.

Table 2 shows how ERDF money is actually being put to use within the Western Isles during 2007-13, as distinct from SPD aspirations.

a) The two largest categories of Western Isles ERDF project funding are schemes to help scattered communities (47.1%) and within-islands accessibility improvement schemes (41.2%). The latter are almost all 'gap filling' transport schemes, which in the Western Isles usually comprise improving link roads to the so-called Central Spine of causeways and bridges running the length of the archipelago;

b) (b) In terms of types of projects, in the Western Isles approximately half are reactive, but mostly in the sense of building on projects started in earlier programs rather than new reactive initiatives in their own right. Again, most of these are 'gap filling' transport schemes. The other half are pro-active, mostly seeking to exploit newer forms of tourism and to a lesser extent media and culture (including the important Gaelic culture of the Western Isles);

c) Renewable energy projects are thin on the ground in the Western Isles and are dominated by local community schemes rather than producing electricity for export. The Western Isles is unfortunate in lacking sufficient undersea cable capacity to exploit its massive renewable energy (offshore and onshore) potential. It will require major infrastructure investment both within the islands and in the form of an undersea cable to the mainland and a cross-Scotland cable to the east coast electricity grid for this potential to be unlocked;

d) Comparing the Western Isles with the wider H&I region, it can be seen that the islands remain more dependent on reactive communications projects and on community level initiatives than the rest of the region. On the other hand, the islands have embraced the need to develop newer sectors and types of businesses on a greater scale than the rest of the region (76.5% to 69.3%). Most of these are in newer forms of tourism (especially activity and cold-water tourism) as the bottom row of Table 2 shows;

e) Most striking of all is the apparent difficulty being faced in the Western Isles in joining what the rest of the H&I region is developing: what are seen as 'key driver' industries (i.e., industrial clusters for the future). We have already seen the problems faced in developing renewable energy for export in the Western Isles. However, other H&I 'key driver' sectors such as Health & Life Sciences and Food & Drink (the latter a particular Scottish success story) are largely missing from the Western Isles. Only in Aquaculture (especially fish farming) do the islands seem to have a comparative advantage. The islands clearly, therefore, are being forced to specialize on a narrower set of niche sectors for the future than other parts of the H&I region. This weakness is also revealed when the ERDF funded HIE schemes omitted from Table 2 are considered. By June 2011, the Western Isles had received business support help under eight separate HIE programs amounting to only 7.6% of all firms assisted in H&I. In contrast, the Inner Moray Firth sub-region around the H&I capital of Inverness, and where many of the regions new growth sectors are based, obtained 32.7% of HEI help. This was not deliberate HIE policy. It simply reflects demand for the schemes.

**Table 2.**A Classification of ERDF Approved Projects in the Western Isles and H&I Region, August 2011. Figures in the table are percentages of the total number of projects in each category.

| *Type of project* | *Western Isles NUTS3 Region* | *Highlands & Islands NUTS2 Region* |
|---|---|---|
| Accessibility – communications within sub-region | 41.2 | 22.7 |
| Accessibility – communications across the wider H&I region | 0.0 | 4.5 |
| Sparse population – general | 0.0 | 0.0 |
| Sparse population – specific scattered communities | 47.1 | 21.6 |
| Islandness (including 'double insularity') | 17.6 | 10.2 |
| Mountain terrain | 0.0 | 0.0 |
| Reactive | | |
| - Infrastructure | 5.9 | 10.2 |
| - Other | 0.0 | 0.2 |
| Reactive, builds on earlier project ('gap filling') | | |
| - Infrastructure | 35.3 | 9.1 |
| - Other | 0.0 | 0.0 |
| Proactive – makes geographical characteristic a positive virtue | | |
| - Infrastructure | 0.0 | 0.0 |
| - Other | 0.0 | 0.0 |
| Proactive – exploits a natural or cultural asset | | |
| - Infrastructure | 23.5 | 28.4 |
| - Other | 17.6 | 29.5 |
| Environmental sustainability (e.g. renewable energy) | | |
| - Infrastructure | 11.8 | 14.8 |
| - Other | 0.0 | 9.1 |
| Sectoral focus – general | | |
| - Traditional sectors | 35.3 | 20.5 |
| - Newer sectors | 76.5 | 69.3 |
| Sectoral focus – key driver sectors | | |
| - Renewable energy (community use) | 5.9 | 21.6 |
| - Renewable energy (export) | 0.0 | 6.8 |
| - New forms of tourism | 35.3 | 43.2 |
| - Aquaculture | 5.9 | 0.0 |
| - Creative and media industries | 5.9 | 8.0 |
| - Textiles and other manufacturing | 0.0 | 2.2 |
| - E-businesses | 0.0 | 4.5 |
| - Life & health sciences | 0.0 | 0.0 |
| - Food & drink | 0.0 | 2.3 |

Note: Figures do not add to 100% since the classification is not mutually exclusive.
Source: Own research.

While the Western Isles continues to struggle to move on from a reactive policy focus, Bornholm had already by 2000-06 taken more radical steps in developing new niches. For its 2007-13 program, Bornholm was faced with a small reduction in its EU budget (down from €8.5m in 2000-06 to €7.5m, or US$11.7m, in 2007-13, at current 2012 exchange rates here and elsewhere where dollar values are given).

There are four priorities in the 2007-13 program: innovation (30% of the budget), utilization of new technology (20%), entrepreneurship (20%) and technical assistance (20%). Bornholm has chosen to align its ERDF spending closely with a new economic development strategy established by the Bornholm Growth Forum set up in 2007. Bornholm's Growth Forum was the first to introduce an industrial cluster emphasis into regional policy in Denmark. Attempts are being made to develop key clusters based around the island's pre-existing strengths (agriculture, food products, light industry & engineering and construction). However, the strategy is more radical than merely protecting established strengths. New niches are being actively sought, with softer types of business support leading the way. The focus is on harnessing new technology to widen and refresh the business base, to re-brand tourism as an 'experience island', and to systematically use policy help to establish Bornholm as a 'green test island' or 'Bright Green Island'. The 'green test island' policy is designed to develop Bornholm as "the setting for a wide range of Danish and international experiments and demonstration projects focusing on electric cars, solar panels, energy-efficient construction and the development of an intelligent electricity system" (Bornholms Regionscommune, 2012). This is one component of a wider strategy introduced in 2008, and designed to be completed in 2014, to fully re-brand all aspects of the island's economy as the 'Bright Green Island'. Eventually, it is hoped that the island will become wholly energy selfsufficient (ibid.). How successful these new policies will be remains to be seen. The strategy does, however, mark a radical step forward from the reactive policies of the past towards proactive and sustainability schemes.

In stark contrast, Lesvos does not seem to have attempted a significant shift in strategic priorities in 2007-13 compared to 2000-06. The need to try to widen the economic base of the islands continued to be stressed in the 2007-13 OP, but in practice the different types of infrastructure investment continue to predominate, as Table 3 shows. The lack of focus on new economic activities and more modern niche market activities is surprising, especially as the key niche sectors (as we have seen in agriculture, food products and tourism) have lost export competitiveness in recent years and even tourism has stagnated. The need to develop new niches is therefore pressing. This does not seem to be well reflected in the program strategy.

Table 3 shows that transport infrastructure spending remains high (32.3% in 2007-13, almost the same as in 2000-06), and heavy spending is continuing in all of the other main infrastructure categories. There has been a small increase in natural heritage projects, but note the drop in the share of help for the cultural industries. The overall picture is therefore one of continued reliance on reactive policies and within this category on hard infrastructure projects.

**Table 3.** Allocations by field of intervention, 2000-06 and 2007-13, Voreio Aigaio region

| | *2000-06 Program* | | *2007-13 Program* | |
|---|---|---|---|---|
| *Field of intervention* | *€m* | *%* | *€m* | *%* |
| Transport | 121.8 | 34.3 | 120.6 | 32.3 |
| Entrepreneurship | 53.9 | 15.2 | 39.2 | 10.5 |
| Communications | 1.6 | 0.5 | 1.9 | 0.5 |
| Protection of the environment | 54.5 | 13.5 | 54.9 | 14.7 |
| Natural heritage | 2.7 | 0.7 | 11.0 | 2.9 |
| Cultural industries | 40.0 | 11.2 | 15.5 | 4.2 |
| Urban regeneration | 6.6 | 1.9 | 28.4 | 7.6 |
| Access of women to work | 20.3 | 5.7 | 14.4 | 3.9 |
| Educational infrastructure | 35.5 | 10.0 | 42.6 | 11.4 |
| Health and social care infrastructure | 21.6 | 6.1 | 44.5 | 11.9 |
| *Total* | *358.5* | *100.0* | *373.0* | *100.0* |

Note: Major changes in Priority names and component elements make it difficult to compare like with like. Reclassification by authors has proved necessary.

# 4. CONCLUSION

The study has produced five main findings. Firstly, it is dangerous to consider islandness in isolation from other geographical challenges to economic development. This is not a new finding, but it is interesting that the case study islands have confirmed that islandness cannot be taken in isolation from other challenges. The interactions of the different geographical challenges can also be very subtle. For example, mountainous terrain was not revealed to be as serious a challenge as the other two EU 'specific geographical features'. Moreover, while sparse population was seen as a serious challenge, it appears to affect regions principally because it is associated with a pattern of scattered small settlements rather than through low population density per se. Islandness is also affected by other geographical features such as the configuration of the landscape and unique 'accidents of geography'. Perhaps the most serious concern facing the case study islands is the demographic challenge of depopulation and an ageing workforce. Given falling birth rates right across the wider EU, it is likely that many other

EU islands also face similar demographic challenges to those of the case study islands.

Secondly, the islands were shown to be heavily dependent on niche market specialization. Some of these were very successful and suggest how islands could succeed in moving into higher value added niches (e.g. Harris Tweed). However, in all three case studies: (a) the existing niches are insufficient to sustain full employment and prosperity, and (b) all have struggled to develop wholly new niches; most have had to rely on consolidating their existing niches.

Thirdly, while ERDF programs proved to be valuable and generated many benefits, the islands continue to be over-focused on reactive policies. These are 'defensive' policies designed to mitigate the geographical challenges (e.g. transport and telecommunications infrastructure to overcome the 'tyranny of distance'). Pro-active policies are growing in number and there is a welcome subtle shift in views from islandness being seen as a problem to being an opportunity. Nevertheless, there is still not enough of this type of policy or of success yet in developing sustainable growth. In this respect, there are huge differences among the three case studies, with Bornholm being much further along this road than Lesvos. The Western Isles are somewhere in between.

Fourthly, the study revealed just how many similarities in strengths, weaknesses, opportunities and threats the islands, mountain regions and SPAs shared. The number of similarities greatly exceeded the differences (e.g. importance of demography and settlement patterns). This is not to argue that the separate analysis of each type of region is unnecessary. On the contrary, there is enormous variation in both the subtleties of the combinations of geographical challenges faced, and the policy strategies appropriate to them. This variation is great within each geographical category. Care must therefore be taken in deriving generalizations and each case should be seen on its own merits.

Finally, the study has shown that, where islands are part of a bigger administrative unit (as is the case with all three islands in this study), additional constraints and limitations can sometimes arise in the manner in which ERDF policies are designed and implemented. The three case study islands each in their own way sought to adapt to the need to work closely in partnership with the larger administrative authority of which they were part. There is wide variation across the EU in how much administrative autonomy particular islands have, but the majority of EU islands do find themselves being part of bigger administrative units and lack the kind of wider autonomy that island states and sub-national island jurisdictions enjoy.

# ACKNOWLEDGEMENTS

The research for this paper was financed by the European Commission in the context of the "Study on the relevance and the effectiveness of ERDF and Cohesion Fund support to Regions with Specific Geographical Features – Islands, Mountainous and Sparsely Populated areas" (Contract: 2010.CE.16.0.AT.059). The views expressed here are those of the authors and do not represent the official views of the European Commission. Moreover, any remaining errors or misunderstandings are the responsibility of the authors. In addition to the European Commission, we thank all the respondents that participated in the research interviews on the three islands of Lesvos, Bornholm and the Western Isles. We also thank two anonymous referees for their most useful comments and suggestions on an earlier draft.

# REFERENCES

1. ADE - Aide à la Décision Économique (2012) Study on Relevance and Effectiveness of ERDF and Cohesion Fund Support to Regions with Specific Geographical Features: Islands, Mountainous and Sparsely Populated Areas, Final Report, Brussels, European Commission, http://ec.europa.eu/regional_policy/sources/docgener/evaluation/pdf/eval2007/geographical_final1.pdf; http://ec.europa.eu/regional_policy/sources/docgener/evaluation/pdf/eval2007/geographical_final2.pdf

2. Armstrong, H.W. & Taylor, J. (2000) Regional Economics and Policy, Oxford, Blackwell

3. Baldacchino, G. (2006) 'Warm Versus Cold Water Tourism: A Review of Policy Implications', Island Studies Journal, Vol. 1, No. 2, pp. 183-200.

4. Baldacchino, G. (2007a) 'Introducing a World of Islands', in G. Baldacchino (ed.) A World of Islands, Malta & Canada, Agenda Academic & University of Prince Edward Island, pp. 1-29.

5. Baldacchino, G. (2007a) 'Introducing a World of Islands', in G. Baldacchino (ed.) A World of Islands, Malta & Canada, Agenda Academic & University of Prince Edward Island, pp. 1-29.

6. Baldacchino, G. & Milne, D. (eds.) (2000) Lessons from the Political Economy of Small Islands: The Resourcefulness of Jurisdiction, Basingstoke, Macmillan.

7. Bardolet, E. & Sheldon, P.J. (2008) 'Tourism in Archipelagos: Hawaii and the Balearics', Annals of Tourism Research, Vol. 35, No. 4, pp. 900-923.

8. Bornholms Regionscommune (2012) www.bornholm.dk

9. Briguglio, L., Cordina, G. & Kisanga, E. (2006) Building the Economic Resilience of Small States, Malta and London, Islands & Small States Institute, University of Malta and Commonwealth Secretariat.

10. Constantino, T.E. & Greene, J.C. (2003) 'Reflections on the Use of Narrative in Evaluation', American Journal of Evaluation, Vol. 21, No. 1, pp. 35-49.

11. Copus, A.K. (1999), A New Peripherality Index for the NUTSIII Regions of the European Union, ERDF Study 98/00/27/130, Brussels, University of Aberdeen/European Commission.

12. Djikstra, L. & Poelman, H. (2008) Remote Rural Regions: How Proximity to a City Influences the Performance of a Region, Brussels, European Commission, Working Paper 01/2008, DG Regional Policy.

13. Djikstra, L. & Poelman, H. (2011) Regional Typologies: A Compilation, Brussels, European Commission DG Regional Policy Working Paper 01/2011.

14. ESPON – European Spatial Planning Observatory Network (2011) Euroislands Project: Draft Final Report, www.espon.eu and http://geoellanikos.aegean.gr/espon/index.php

15. Eurisles (2002) Off the Coast of Europe: European Construction & the Problem of the Islands, Ajaccio, Corsica, for Islands Commission, Conference of Peripheral & Maritime Regions.

16. European Commission (1996) The Structural Funds and Cohesion Fund 1994-99: Regulations and Commentary, Luxembourg, Office for Official Publications of the European Communities.

17. European Commission (2006) Regions & Cities for Growth and Jobs: Overview of Regulations 2007-13 in Cohesion & Regional Policy, Brussels, Inforegio Factsheet, DG Regional Policy.

18. European Commission (2012a) The Outermost Regions: Regions of Assets and Opportunities, Brussels, Regional Factsheet, DG Regional Policy.

19. European Commission (2012b) EVALSED: Evaluation and Socio-Economic Development, European Commission, DG Regional Policy, Brussels, http://ec.europa.eu/regional_policy/information/evaluations

20. Eurostat (1994) Portrait of the Islands, Luxembourg, Office for Official Publications of the European Communities.

21. Flyvbjerg, B. (2006) 'Five Misunderstandings about Case Study Research', Qualitative Inquiry, Vol. 12, No. 2, pp. 219-245.

22. Gløersen, E., Dubois, A., Copus, A. & Schürmann, C. (2006) Northern Peripheral, Sparsely Populated Regions in the European Union & Norway, Report 2006/2, Stockholm, Nordregio.

23. Hall Aitken (2007) Outer Hebrides Migration Study: Final Report, Report for Comhairle nan Eilean Siar (council), Stornoway, Western Isles Enterprise & Communities Scotland.

24. Highlands & Islands Enterprise (2009) Area Profiles: The Western Isles, Inverness, Highlands & Islands Enterprise.

25. Instituto per la Ricerca Sociale (2011) Study on the Contribution of Local Development in Delivering Interventions Co-Financed by European Regional Development Fund in 2000-06 & 2007-13: Final Report, Brussels, Study for DG Regional Policy of the European Commission.

26. Koenig, G. (2009) 'Realistic Evaluation and Case Studies: Stretching the Potential', Evaluation, Vol. 15, No. 1, pp. 9-30.

27. Martini, A. (2009) Making Impact Evaluations more 'Rigorous' through Use of Counterfactual Impact Evaluation Methods, European Commission, 6th Evaluation Conference, Warsaw, http://ec.europa.eu/regional_policy/archive/conferences

28. Moncada, S, Camilleri, M., Formosa, S. & Galea, R. (2010) 'From Incremental to Comprehensive: Towards Island-friendly European Union Policy-making', Island Studies Journal, Vol. 5, No. 1, pp. 61-68.

29. Monfort, P. (2009) Territories with Specific Geographical Features, Regional Policy Working Papers, No. 02/2009, DG Regional Policy, Brussels, European Commission.

30. Planistat Europe (2006) Analysis of Island Regions and Outermost Regions of the European Union, Report No. 2000.CE.0.AT.118, DG Regional Policy, Brussels, European Commission.

31. Samsø Energy Academy (2007) A Renewable Energy Island: Ten Years of Development, Samsø, Denmark Samsø Energy Academy.

32. Snedden Economics (2007) Outer Hebrides Tourism Update, Stornoway, Report for Highlands & Islands Enterprise, VisitScotland(Outer Hebrides) & Comhairle nan Eilean Siar

33. Technopolis Group (2012), Evaluation of Innovation Activities: Methods and Practice, Study for DG Regional Policy of the European Commission, Brussels.

34. Yin, R.P. (1981) 'The Case Study Crisis: Some Answers', Administrative Science Quarterly, Vol. 26, No. 1, pp.58-65.

8

# GAZING AT HAW PAR VILLA

**Voon Chin Phua , Joseph W. Miller**

Gettysburg College

## ABSTRACT

Tourism is an important and growing industry in Singapore. Studies on Singapore cultural tourism have generally focused on three major sites: Chinatown, Little India and the Malay Village. The Haw Par Villa tourist site has not been examined in recent years. The case study of Haw Par Villa offered here demonstrates how changing times in Singapore have affected the popularity of tourist sites in an island nation. This article discusses the decline and potential rebirth of Singapore's Haw Par Villa theme park in the context of cultural tourism, placing a special emphasis on Urry's concept of the 'tourist gaze'. Multiple methods were used in gathering data for this study: a survey conducted in Singapore of both local residents and foreign tourists; participant observations of Haw Par Villa; and a thematic content analysis of tour guide books and online documents pertaining to the site. Our analyses suggest that Haw Par Villa represents a treasured past of Singapore, although one in danger of fading away with the changing interests of newer generations of tourists.

## KEYWORDS

Tourism, Singapore, tourist gaze, cultural tourism

# 1. INTRODUCTION

Tourism is an important revenue-generating industry for some islands and island nations (see Royle, 2009) whose competitiveness is dependent on their "size, structure, and quality of tourist attractions, infrastructures and super structures, and the destination's management capabilities" (Croes, 2011: 432). In Singapore, tourism is an important and growing industry (Ramchandani, 2012). According to the Singapore Tourism Board (STB, 2013), Singapore received S$23 billion in tourism receipts in 2012, which is a 3% increase from 2011 and more than three times higher when compared to the amount in 2003.

Studies on Singapore cultural tourism generally have focused on three major sites: Chinatown, Little India, and the Malay Village (eg Phua, Berkowitz and Gagermeier, 2012; Chang, 2000). In the late 1980s these three areas were earmarked as historical sites, representing three distinct and major ethnic groups in Singapore. Other cultural sites, like the Chinese and Japanese gardens and Haw Par Villa, are some distance from popular tourist destinations and require a greater sense of determination for tourists to visit them. Among these sites, Haw Par Villa is of particular interest because of its history and its dramatic rise and fall as a tourist site that features Chinese culture. The promotion of particular cultural sites may vary over time as newer and more popular sites are established. For example, the recent opening of casinos in Singapore is a way to attract new and different tourists, although the casinos are not without their controversies (eg Ooi, 2005). This case study demonstrates how changing times in Singapore have affected the popularity of tourist sites in an island nation.

In this article, we examine the current state of Haw Par Villa as a tourist attraction. In order to do that, we first provide an overview of Singapore tourism in recent years and a brief history of the development of Haw Par Villa. Then, we describe our data and methodology. Finally, we present our analysis and discuss its implications. In our discussion, we highlight the complexities of Haw Par Villa as a tourist site in Singapore.

# 2. Tourism in Singapore

Singapore is an island nation located at the gateway of Southeast Asia. Its 2010 census reports that the population is currently about five million, of which are 26% are nonresidents (Singapore Department of Statistics, 2010). 74% of its population self reported as Chinese, 13%

Malay, 9% Indians and 3% Others. For this small nation, tourism is an important source of income. Singapore's emergence in the international tourism sector is relatively recent, with tourism only registering on the government's radar in the mid 1960s after Singapore's economic value as an entrepôt and industrial city began to wane. During this time, tourism was seen as "an avenue to generate employment and enhance infrastructural development" (Chang, Milne, Fallon, and Pohlmann, 1996: 291-294). Along with the obvious economic value, though, Ooi argues that the tourism industry "played a central role in shaping Singaporeans' own understanding of their national and ethnic identities" (2005: 250). Keeping up with the times, STB reports that, "[f]or Singapore to continue to excite and appeal, she has to constantly evolve to present new and engaging travel experiences" (2010: online).

Singapore's tourism brand has gone through several iterations in the past few decades in response to changing tourist demands and as a reflection of the country's continued diversification and commitment. Over the years, STB has been leveraging the country's limited natural resources and short history to promote Singapore as a tourist destination. In earlier years, Singapore designated several places as ethnic cultural sites, serving both to promote cultural diversity and tourism. 2010 saw the launch of Singapore's casino resorts as major tourist sites. STB pledges "to lure 17 million visitors to Singapore and triple tourism revenue to S$30 billion by 2015" (Bennett and Chen, 2009).

According to the STB Overseas Visitor Survey (2013), the most visited paid-access venues among tourists were MBS Skypark, Night Safari, Sentosa, Singapore Flyer and Universal Studios Singapore; while the main sites with no admission fee were Chinatown, Little India, Marina Bay, Merlion Park and Orchard Road. Once hailed as constituting "a rich heritage of Chinese culture" (Teo and Yeoh, 1997: 199), Haw Par Villa used to be a very popular tourist attraction. In 1984, Haw Par Villa was the fourth most popular tourist attraction in Singapore, visited by 31.3% of all visitors to Singapore

(Nirman, 2000: 53). However, Haw Par Villa lost its popularity among both local and foreign visitors from the mid-1980s. Clearly, Haw Par Villa is now no longer a major tourist attraction, not making the list of top tourist attractions on sites promoting Singapore tourism (eg World Tourist Attractions, 2011; www.singapore-vacationattractions.com, 2013).

# 3. THE HISTORY OF HAW PAR VILLA

Tiger Balm Gardens, commonly referred to as 'Haw Par Villa', is a nearly century-old theme park located on the south western side of Singapore. The theme park was created by Chinese businessman and philanthropist Aw Boon Haw, who created and marketed the wildly successful Tiger Balm with his younger brother Aw Boon Par. The villa residing within the gardens was actually Haw's gift to his business partner and brother, Aw Boon Par. Construction of the villa and the surrounding gardens began in 1935 at a staggering cost of US$1.95 million. While the majority of the permanent installations were completed by 1937, Aw Boon Haw's nephew (Renuka, 2002) or son (Brandel and Turbeville, 1998), depending on the source, continued to add and maintain the gardens even after Haw's death in 1954.

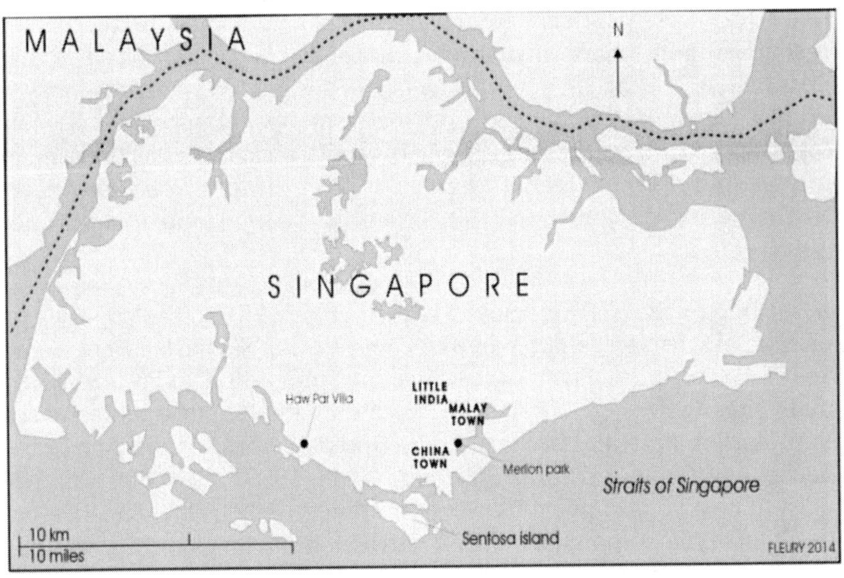


**Figure 1.** Map of Haw Par Villa (Christian Fleury, 2014)

Aw Boon Haw's intent in creating the public gardens remains unclear. Some historians conclude that his primary motivation was a concern that Chinese immigrants would forget their own culture after living overseas and mingling with westerners. Therefore, as the story goes, he included morally educative tableaux depicting cherished traditional stories in order to maintain a sense of Chinese national heritage. Still others believe the park was yet another one of his clever advertising schemes — this idea being largely derived from the park's namesake product, the Tiger Balm heat rub (Brandel and Turbeville, 1998). Regardless of his intent, the park undeniably emphasises Chinese myths, legends and Confucius' teachings.

**Figure 2.** The majestic entrance to Haw Par Villa theme park (photograph by Voon Chin Phua, 2011)

The content of the gardens tends to evoke several different reactions. Renuka comments on the garden's ability to convey traditional Chinese stories in a "modern, Western way" (2002: online). Brandel and Turbeville reflect this idea by commenting that the gardens "mingle traditional and modern, Eastern and Western motifs" (1998: 109). The latter may largely be a result of Haw's son including the 'International Corners' after his father's death, including but not limited to Japanese sumo wrestlers and a version of the Statue of Liberty with an Asian-looking face. Straying from the style of the depictions, Brandel and Turbeville further comment that the gardens "delight, soothe, and frighten in just the right proportions" (1998: 14). These varying sentiments have been captured in several websites and blogs commenting on the eccentric Singaporean theme park, although recent commentary has tended to accentuate the park's gruesome and horrific elements such as the Ten Courts of Hell (Discario, 2010; Hemmy.net, 2008).

## 4. THE RISE AND FALL OF HAW PAR VILLA AS A TOURIST SITE

In its early days, Haw Par Villa was a popular and free public park. Teo and Yeoh report that people visited Haw Par Villa to "[learn] from and [appreciate] the tableaux on Chinese history, mythology, and folklore which they took to represent the heritage of the majority ethnic group in Singapore" (1997: 204) as well as to use it as a leisure spot. In STB's attempt to salvage "the loss of the island's oriental mystique" in the 1980s, Haw Par Villa became:

> a showcase of Chinese heritage [as it] fell under the 'Exotic East' theme. The board's recommendation was to turn the park into a 'high-tech entertainment' center... while at the same time adapting modern technology for Asian culture to create a Chinese mythology park. (Teo and Yeah, 1997: 199)

The STB outsourced the park's operation to an international corporation, which proceeded to commercialise the park, adding live attractions and theme-park rides.

The park's commercialisation was a resounding failure, causing losses in the millions of dollars (Renuka, 2002). One reason for its decline was that the static presentations of mythological events had become outdated. Younger generations now expect entertainment that is more technologically sophisticated. In addition, younger generations may also have become less familiar with the mythological events and Confucius' teachings as the educational system in Singapore changed over the years (Lim, 2004). While attempts were made to upgrade Haw Par Villa to a technology- and capital-intensive theme park; the provision of live performances and adventure rides created high operational costs that could only be covered by charging an admission fee. Singapore simply did not have the critical tourist mass that could afford to pay such high admission fees. Teo and Yeoh (2001) posit that the hiring of foreign consultants who did not consider local conditions resulted in miscalculations of costs and revenues. As a result, the STB resumed operation and returned the park to its original status as a public amenity. "Since STB took responsibility for the park on April 1, 2001, there has been a threefold increase in visitors. More than 70 percent of visitors to the park are tourists, primarily from China, Hong Kong, Indonesia, Malaysia and Taiwan" (STB Annual Report 2001-2002). While the glory days of packed sidewalks are probably behind the park forever, several writers and bloggers have expressed appreciation for the return of the park as a public venue, thus more clearly reflecting its heritage as a philanthropic gift from a cherished resident (Brandel and Turbeville, 1998).

## 5. THE TOURIST GAZE AND HAW PAR VILLA

There is little systematic reporting on the extent to which Haw Par Villa has changed over the past decade and how it currently stands as a tourist attraction. This lack of interest indicates that Haw Par Villa is no longer a popular tourist destination and that tourists' tastes have changed. Nevertheless, Haw Par Villa remains a historical cultural landmark in Singapore. In this article, we are interested the current views of Haw Par Villa as a tourist attraction in this island nation where new tourist sites are constantly being developed to capture more tourists in this ever-changing industry. In Urry's (1990a) terms, should tourists visit to gaze at Haw Par Villa? Urry (1990a) developed the concept of tourist gaze in his analysis of tourism:

What makes a tourist gaze depends upon what it is contrasted with, what the forms of non-tourist experience happen to be. The gaze therefore presupposes a system of social activities and signs which locate the particular tourist practises, not in terms of some intrinsic characteristics, but through the contrasts implied with non-tourist social practises, particularly those based within the home and paid work. (1990a: 2)

In addition, Urry (1990a: 144) argues that "If an ethnic group living in an area is sometimes constructed and presented as a visitor attraction, another way of expressing this is to say that ethnic differences, the 'exotic', may sometimes function as a 'theme'" (ibid: 144). Examining the case of Haw Par Villa, we can argue that representations of Chinese culture are considered a thematic attraction for non-Chinese tourists. Even for Chinese tourists, the idea of overseas Chinese renditions of Chinese cultures may appear foreign and thus are both familiar and exotic at the same time. Anticipation is important in choosing a particular site (sight) for the tourist's gaze: "[s]uch anticipation is constructed and sustained through a variety of non-tourist practices... which construct the gaze" (Urry, 1990b: 26).

Several studies have used this concept in framing their research. Bai argues that, cultural traditions "have become mere raw materials in the construction of ethnic identities, which are mediated by official terminology as well as by the effects of the tourist gaze" (2007: 225). However, Bai also posits that the gaze is not unidirectional, "It presents a chance for domestic tourists to gaze at foreign tourists and consume a foreign atmosphere on the Chinese territory" (2007: 253). Moufakkir (2011) reverses the concept and studies how cultural distance affects Dutch hosts' views on German and Asian tourists. Researching solo female tourists, Jordan and Aitchison (2008) suggest that the mutual gaze and the local gaze are as critical as the tourist gaze in understanding gendered power between tourists and the locals. As a constructive critique and further expansion of Urry's tourist gaze, Perkins and Thorns (2001) argue that tourist performance may better capture more diverse involvements among tourists instead of focusing primarily on the visual. In her study of food tourism, Everett posits that it "provides an interesting embodied dimension to the way that the 'gaze' is conceptualized; it is an experience that can only be pursued if one goes beyond visualism" (2008: 345). MacCannell (2001) argues that the tourist gaze is conceptually deterministic even though it focuses on the consumers instead of the producers and proffers the idea of the second gaze "that is capable of recognizing the misrecognition that defines the tourist gaze" (2001: 30). The second gaze "may be more interested in the ways attractions are presented than in the attractions themselves" on tours (ibid: 36).

In the latest rendition of tourist gaze, Urry and Larsen have expanded the concept of the gaze to include service work and rethought "the tourist gaze in the light of this performance turn and of a broader Goffmanian dramaturgical sociology" (2011: 189). Incorporating the idea of performance turn in tourist gaze, they attempted to address the criticisms of tourist gaze generated from its inception more than ten years ago. "The performance turn highlights how tourists experience places in multi-sensuous ways that involve bodily sensations and affect" (Urry and Larsen, 2011: 190). However, while they incorporated new ideas and expanded how the visual is an important aspect of tourism, they retain the gaze as their central approach in examining and understanding tourism.

In this article, we uses the concept of the tourist gaze to develop an analytic framework where we specifically examine how Haw Par Villa is discussed and constructed as a tourist attraction. This is an important question in the context of land-scarce island nations where profitable land use sometimes conflict with cultural and historical preservation. However, it is important to note that if we take it literally, the concept of the tourist gaze seems to favour the dominance of the visual sense over other sensory engagements (Urry, 1992a, 2001). While the increasing growth of complex tourism practices and experiences means a visual approach is inadequate to address how space is experienced multi-dimensionally (eg Everett, 2010); this concept suffices as the size and the services provided at Haw Par Villa cater mainly to the visual sense over other sensory engagements.

## 6. METHODS

In researching this article, we used multiple approaches to collect our data to ensure reliability. Using Google as the search engine in 2009-2010, we collected data from the Internet on Haw Par Villa that consisted mainly of newspaper articles, blogs and tourist information sites. We also used information from a survey conducted in the summer of 2009 on cultural tourism in Singapore. The survey included both tourists and locals and was conducted by three researchers. In order to find participants for our survey we frequented several tourist destinations and pro-actively solicited interviews. Our sample size was 147. Our participants consisted of 49% men and 51% women. Their ages ranged from 18 to 70 years old. 41% of them were first-time visitors to Singapore, 27% had visited Singapore before, and 32% were native Singaporeans. 94% of interviewees were non-Asian tourists; only 6% were Asian visitors. The two countries most represented in our sample were the United Kingdom (16%) and Australia (14%). Other countries include USA, Germany, Nepal, Norway, Sri Lanka, Argentina, Hungary, Russia, Indonesia, Slovakia and Trinidad and Tobago. In addition, we visited the Haw Par Villa in the summer of 2009 and 2010, writing field notes and photographing the site. Finally, we reviewed 25 tourist guidebooks from the in-store stock of large, chain bookstores in the US (February-May 2009) and Singapore (June-July 2009) to examine whether and how Haw Par Villa was being promoted.

Thematic content analysis was conducted to examine how individuals discuss Haw Par Villa as a tourist site. We used a grounded theory approach in analysing the data (Glaser and Straus, 1967), allowing themes and issues that participants deemed important to emerge rather than to structure the analysis using predefined frameworks. First, we performed open coding of broad themes, followed by focused coding using those identified themes (eg Esterberg, 2002; Charmaz, 2006). Each set of data was analysed separately but were consolidated to examine whether the views from each source were similar to or contradict one another.

## 6.1 Land of the Forgotten Past

Of the tourists surveyed in 2009, no one mentioned Haw Par Villa as one of their top three places to visit. For most tourists, the top sites in Singapore that they wanted to visit included Sentosa (a beach resort island), the Night Safari, shopping malls (specifically Orchard Road), the Singapore Flyer, Chinatown, Little India and the Botanic Gardens. Among native Singaporean survey participants, only three mentioned Haw Par Villa as a unique place to visit in Singapore. The main reasons given for recommending Haw Par Villa were its roots in Chinese myths and traditions, its proximity to the Hua Song Museum and its unique combination of Eastern and Western statues.

On travel advice blogs, the most commonly mentioned attraction within Haw Par Villa is the Ten Courts of Hell, which depicts the punishments for sins committed while one is alive. 'Spooky' describes it thus:

According to Chinese mythology, a soul must pass through ten courts of judgment, before being allowed into heaven, and Haw Par Villa has very detailed representations of these trials. (2010: online)

Three Australians and one American on www.tripadvisor.com (2010-2011) share their views: "I must add that the 'Ten Courts of Hell' exhibit is not be missed... take your kids - trust me - they'll never be naughty again." (Australia 2009); "The Ten Courts of Hell... features the ten steps of judgment before reincarnation... The statues and sets depict moral values and the punishments for wrong-doing. (Australia 2011); and "We did have to pay $1 to view the Ten Courts of Hell but it was well worth the $1". (USA, 2006). However, one Australian (2010) did not think the site was worth visiting: "What were they thinking when they designed this place for their children? The 12 paths to hell! It was like a puppet show but gory. It was tacky and not worth the entry fee!" Of great interest is the qualification that the place is rather rundown and not very well maintained. For example, Selanan (www.tripadvisor.com) from the UK mentions that:

**Figure 3** .A sinner's heart is carved out in the Ten Courts of Hell (photograph by Joseph W. Miller, 2011)

The whole attraction is obviously in need of a cash injection and a marketing revamp, such a shame that it has been left to stagnate as it could be so much more than it is at the moment. The statues are fascinating, the folklore riveting, loads of space, vibrant colours, fabulous photo opportunities.... but the atmosphere is of an old run-down theme park with very little going on. (2011: online)

Ironically, for one Singaporean survey participant, this provides an impetus to see it before it is no longer there. Another Singaporean survey participant actually commented that, "it is there but forgotten, like the land of the forgotten past. The younger generation has no interest in it and the older has no time for it. Who knows if the government will tear it down sooner or later" (2011: online).

Figure 4 – Statues like this with human heads are scattered throughout the park (photograph by Joseph W. Miller, 2010)

## 6.2 Walking down Memory Lane

Of the 25 guidebooks we reviewed, only one had a substantive write-up on Haw Par Villa: "[t]here is hardly a native resident in Singapore who hasn't been taken here as a child. For generations parents have brought their children here to expose them to traditional values – mostly Confucian – behind these stones" (Brown, Backenheimer and Tan, 2009: 64-65). This statement is probably true of the generations growing up in the 1960s-1980s. As a result, for many adult Singaporeans, Haw Par Villa represents part of their childhood memory. On his blog entitled "Haw Par Villa Re-visit", Laokokok (2007) lamented the changing nature of this site. A notable feature of the blog is two photos of the author, one taken in the 1970s and another in 2007, both standing in front of the same statue, which had been moved in the interim. Responses to the blog included memories of the peak tourist times of the site, with some commenting on the younger generations' lack of appreciation for these Chinese legends and values.

Haw Par Villa brings back childhood memories for many adult Singaporeans. As such, many of them see Haw Par Villa as a part of Singapore's history (Teo and Yeoh, 1997). An important distinction here is the generational divide with respect to how Haw Par Villa is viewed. Younger generations seem mainly to view Haw Par Villa as an outmoded tourist attraction while the older generations see visiting Haw Par Villa as a walk down memory lane. The somewhat rundown look of the park ironically more accurately represents its long period of existence and its decline as a tourist site, as opposed to complaints about the artificiality of the renovated Chinatown. (Oon reports that as "part of the government's heritage conservation mandate, Chinatown was cleaned up and restored, too clinically some may argue, robbing it of some of its authenticity" [2008: 53]).

## 6.3 TO GAZE OR NOT TO GAZE

Haw Par Villa receives mixed reviews as a tourist site. On Lonely Planet's blog, one traveller inquires about the place: "Been told about Haw Par Villa in west of Singapore and told it's quite an experience of Chinese folklore! Anyone been who could advise best way by public transport to get there (and how long you'd suggest spending there)". Responses to this inquiry are mixed; while some simply provided directions, others offered suggestions:

Don't forget to see the Hua Song museum there and also the chamber of horrors.

You won't have to spend more than a couple of hours there as it is quite small. The place has been out of operation for many years so entry is now free, you can still explore the interesting statues there and look around the park however.

Haw Par Villa is very very old and very very run down.... Chances are u'd be extremely disappointed. It's a little kitchy too. Go to Chinatown and some of the Chinese temples there, u can get a better real life experience. (Lonely Planet, 2010)

On our visit to Haw Par Villa, we noticed that many statues were in need of restoration. However, the park itself is a museum piece. The place felt like a walk back in time. Such displays of Chinese folklore are uncommon in Singapore, and the contrast between the newer statues (eg the Statue of Liberty with an Asian face) and the original statues (of Chinese folklore figures) marks an interesting blend of the East meeting the West. Hua Song Museum is a valuable addition, although it has an admission fee. Knowing what to expect is a critical element to appreciating Haw Par Villa.

## 6.4 Gazing at Haw Par Villa

Gazing at Haw Par Villa is not necessarily just looking at something out-of-the-ordinary. The idea that Haw Par Villa is a place of nostalgia for some tourists suggests that Haw Par Villa encompasses elements of the ordinary and the out-of-the-ordinary. The former suggests that comfort and familiarity can be reasons for attraction and the latter denotes that the factor of time could reshape something ordinary in the past into being out-of-the-ordinary in the present. What is also interesting is that such a gaze that involves nostalgia could result in wanting to preserve that place as suggested in our study. This idea is similar to Urry's (1992b) study on the relationship between the tourist gaze and the environment. Urry writes that, "the romantic tourist gaze... feeds into and supports attempts to protect the environment" in spite of the fact that tourism may have adverse effects on the environment (ibid: 9).

What is hidden behind the nostalgia felt by some participants (on websites and in interviews) is akin to the concept of the second gaze. What these participants did not implicitly and directly voice beyond their nostalgia is the changing tastes of tourists and how the current state of Haw Par Villa is a reflection of that. In commenting on Haw Par Villa, participants either comment on what it is not as a viable tourist site or what it was (reliving the past). MacCannell (2001) argues that the second gaze looks at the way attractions are presented than the attractions themselves, though in this case, the current state of Haw Par Villa is not necessarily STB's intended presentation.

## 6.5 The Future of Haw Par Villa

Examining the data sources we collected, tourists are not familiar with or particularly interested in Haw Par Villa as a tourist site and guidebooks do not have extensive writeups on it. During our visits to Haw Par Villa we did not see many tourists (although this is highly dependent on the timing). Not surprisingly, most information presented in this article derives from Internet sources as our other research yielded limited information. However, this lack of information on Haw Par Villa underscores an important point – the lack of promotion and marketing of this site. Thus, the fate of Haw Par Villa lies in part in the priorities of STB, which has a strong influence over the promotion of tourist sites. However, one also has to recognise that Singapore's tourism is dependent on the everchanging preferences of tourists. In the context of an island nation with limited land area, resources may have to be disproportionally allocated to tourist sites that are more able to generate profit (eg Ramchandani, 2012). Haw Par Villa in its current state is no competition to other tourist sites, such as Sentosa and the casinos (eg Henderson, 2006), in terms of its type and variety of entertainment, particularly for the younger generation. Yet Haw Par Villa represents, in its own way, a treasured past, although one in danger of fading away with newer generations of tourists. Managing expectations is an important aspect to securing Haw Par Villa's future. Phua et al (2009) argue the importance of being consistent between what tourist sites offer and how they are promoted. Providing critical information on Haw Par Villa's historical significance (better and clearer descriptive signs for the statues, for instance) and an honest account of its current state in promotional materials will help manage tourists' expectations. According to one visitor:

just because this place lost its glow doesn't mean it's not worth a visit, especially if you're into offbeat attractions. I mean this place will really blow your mind, with statues like a woman breastfeeding her father in law, armed monkeys, or the human faced giant crab. Now these apparently do make some sense if you've read up on Chinese mythology, but to an uneducated guy like me they just look freaky. (Spooky, 2010: online)

## 7. CONCLUSION

Haw Par Villa represents a part of Singapore history and a part of Singapore heritage. The characterisations in this article resonate with similar sentiments in terms of Haw Par Villa being a place of nostalgia for the older generations who had frequented the site and knew more about its history. It is also one of very few places steeped in Chinese folklore and legends in Singapore, inculcating Chinese and Confucian values. In fact, one guidebook hails it as "[f]ar livelier than the Chinese and Japanese gardens ... [and] the most Chinese of Singapore's theme parks" (Brown, Backenheimer and Tan, 2009: 65).

Using Urry's theoretical lens of the tourist's gaze enables us to evaluate how this site is being viewed by visitors (both foreign tourists and locals). While sight (gaze) is one of many senses that tourists could employ in their experience, its importance does not necessary diminish. In fact, one could argue that for some tourists, the sight captured (especially) via a camera is the best evidence they have to show others that they have travelled (Urry and Larsen, 2011). In the context of Haw Par Villa, with all its oddities, is a perfect example of a tourist's visual consumption. Yet, what tourists fail to voice is the second gaze, MacCannell's (2001) critique of the first version of the tourist gaze. The second gaze would enable tourists to look at what the current state of Haw Par Villa is representing.

Also of interest is that people who argue for the preservation of Haw Par Villa are not necessarily local Singaporeans. On the internet, the groups interested in keeping Haw Par Villa open appear rather diverse. However, we are unable to ascertain the precise extent of the diversity and magnitude of this tendency. Nonetheless, the results show that people who are interested in keeping Haw Par Villa open, regardless of their country of origin, share at least one of the following two characteristics: interest in Chinese culture and/or past visitation. As such, Haw Par Villa could be considered as a site for cultural contact where both foreigners and locals visit the place because of its local cultural value. Yet it is also the changing tastes of foreigners and locals that are impacting its popularity, leading to several of its manifestations.

To celebrate the country's fity years of tourism in 2014, STB provided free guided tours to Haw Par Villa to enable "Singaporeans and residents here to relive their memories at local attractions and create new ones with their friends and families" (Chow, 2014: online). Visiting the site in earlier 2014, the authors also noticed some areas were renovated with fresh layers of paint. Still, Haw Par Villa as a tourist sites pales in comparison with the newer and more technologically enhanced sites. The fate of Haw Par Villa as a tourist attraction is interesting as it brings out the tension between the preservation of heritage attractions and generating tourist receipts. Future research could include a study of STB priorities in the challenging landscape of changing tourists' preferences and expectations. Another topic that could be explored is the issue of authenticity. This could be approached in two ways. First, one authenticity issue here is the extent to which Singapore Chinese culture has evolved from the original Chinese culture and its representation (eg Jennings, 2010, on 'inherited' cultural heritage; Hayward, 2012, on evolving representations of Singapore's Merlion symbol). Second, the issue of authenticity and place branding is worth exploring if place branding is used to revive the site (eg Ooi and Stöber, 2010).

## BIBLIOGRAPHY

1. Bai, ZhiHong (2007) 'Ethnic Identities under the Tourist Gaze', Asian Ethnicity v8 n3: 245-259

2. Bennett, S and Chen, S (2009) 'Singapore Expects 'Challenging Year' for Tourism (Update 1)', Bloomberg, January 10th: www.bloomberg.com/apps/news?pid=20601080&refer=asia&sid=asQmaaFwIl1g - accessed May 2009

3. Brandel, J and Turbeville, T (1998) Tiger Balm Gardens: A Chinese Billionaire's Fantasy Environments, Hong Kong: South China Printing, Ltd

4. Brown, J.D, Backenheimer, M and Tan, W.L (2009) Singapore Pocket Guide (fifth edition), Singapore: Berlitz Publishing

5. Chang, T.C (2000) 'Singapore's Little India: A Tourist Attraction as a Contested Landscape', Urban Studies v37 n2: 343-366

6. Chang, T.C, Milne, S, Fallon, D and Pohlmann, C (1996) 'Urban Heritage Tourism: The Global-Local Nexus', Annals of Tourism Research v23 n2: 281-303

7. Charmaz, K (2006) Constructing Grounded Theory: A Practical Guide through Qualitative Analysis, London: Sage Publications

8. Chow, J (2014) 'Free Guided Tours for Local Residents to Mark the 50 years of Tourism', The Straits Times, March 4th, online at: www.straitstimes.com/breakingnews/singapore/story/free-guided-tours-local-residents-mark-50-years-tourism- 20140304 - accessed June 2014.

9. Croes, R (2011) 'Measuring and Explaining Competitiveness in the Context of Small Island Destinations', Journal of Travel Research v50 n4: 431-442

10. Discario (2010) 'Singapore Haw Par Villa Theme Park', Discario: Modern and New Magazine: http://www.Dyscario.com/travel-and-places/singapore-haw-par-villa-themepark.html - accessed September 2010

11. Esterberg, K.G (2002) Qualitative Methods in Social Research, Boston: McGraw-Hill

12. Everett, S (2008) 'Beyond the Visual Gaze? The Pursuit of an Embodied Experience through Food Tourism', Tourist Studies v8 n3: 337-358

13. Hayward, P (2012) 'Merlionicity: The Twenty First Century Elaboration of a Singaporean Symbol', Journal of Marine and Island Cultures v1: 113-125

14. Hemmy.net (2008) 'Singapore Haw Par Villa Theme Park': www.hemmy.net/2008/08/02/Singapore-haw-par-villa-theme-park/ - accessed September 2010

15. Henderson, J (2006) 'Betting on Casino Tourism in Asia: Singapore's Integrated Resorts', Tourism Review International v10 n3: 169-179

16. Jennings, A (2010) 'The Giantess as a Metaphor for Shetland's Cultural History', Shima: The International Journal of Research into Island Cultures v4 n2: 1-14

17. Jordan, F and Aitchison, C (2008) 'Tourism and the Sexualisation of the Gaze: Solo Female Tourists' Experiences of Gendered Power, Surveillance and Embodiment', Leisure Studies v27 n3: 329-349

18. Laokokok (2007) 'Haw par Villa Re-visit': Times of My life website: timesofmylife.wordpress.com/2007/07/13/haw-par-villa-re-revisit/ - accessed September 2010

19. Lim, C.T (2004) 'The Confucian Tradition and its Future in Singapore: Historical, Cultural and Educational Perspectives', in Yong, M.C (ed) Asian Traditions and Modernization: Perspectives from Singapore, London: Eastern Universities Press: 177-207

20. Lonely Planet (2010) 'Thorn Tree Forum – Haw Par Villa', online at: www.lonelyplanet.com/thorntree/forums/asia-south-east-asia-islandspeninsula/Singapore/haw-par-villa?page=1 - accessed August 204

21. MacCannell, D (2001) 'Tourist Agency', Tourist Studies v1 n1: 23-37

22. Moufakkir, O (2011) 'The Role of Cultural Distance in Mediating the Host Gaze', Tourist Studies v11 n1: 73-89

23. Nirman (2000) 'Balm for Ailing Gardens?' The Strait Times (Review Focus), September 3rd, archived online at: eresources.nlb.gov.sg/newspapers/Digitised/Article/straitstimes20000903.2.68.1.aspx - accessed July 2014d

24. Ooi, C.S (2005) 'State-Civil Society Relations and Tourism: Singaporeanising Tourists, Touristifying Singapore', SOJOURN v20 n2: 249-272

25. Ooi, C.S and Stöber, B (2010) 'Authenticity and Place Branding: The Arts and Culture in Branding Berlin and Singapore', in Knudsen, B.T and Waade, A.M (eds) Re-Investing Authenticity: Tourism, Place and Emotions, Bristol: Channel View Publications: 66-79

26. Oon, H (2008) Globetrotter Island Guide: Singapore London: New Holland Publishers Ltd.

27. Perkins, H.C and Thorns, D.C (2001) 'Gazing or Performing? Reflections on Urry's Tourist Gaze in the Context of Contemporary Experience in the Antipodes', International Sociology v16 n2: 185-204

28. Phua, V.C, Berkowitz, D and Gagermeier, M (2009) 'Multicultural Tourism in Singapore', paper presented at the 3rd International Colloquium on Tourism and Leisure, Bangkok, Thailand, July 2009

29. (2012) 'Promoting Multicultural Tourism in Singapore', Annals of Tourism Research v39 n2: 1255-1257

30. Ramchandani, N (2012) 'Welcome to S'pore – but We're Getting Picky', AsiaOne News April 22nd: www.asiaone.com/print/News/Latest%2BNews/Singapore/Story/A1Story20120422- 341210.html - accessed September 2013

31. Renuka, M (2002) 'Tiger Balm Gardens', National Library Board Singapore, archived online at: infopedia.nl.sg/articles/SIP_ 560_ 2004-12-14.html - accessed March 2011

32. Royle, S.A (2009) 'Tourism Changes on a Mediterranean Island: Experiences from Mallorca', Island Studies Journal v4 n2: 225-240

33. Singapore Tourism Board (2001-2002) 'Prickly Projects Show a New Cultural Sensitivity', online at: appstg.stb.gov.sg/Data/pages/1/c4015b7f368e6b4871809f49debe0579/A%20Prickly%20Projects%20Show%20A%20New%20Cultural%20Sensitivity.pdf - accessed September 2011

34. (2010) Brand Overview: Destination Singapore, online at: app.stb.gov.sg/asp/des/des05.asp - accessed May 2011

35. (2013) Singapore's Tourism Sector Performance for 2012, online at: app.stb.gov.sg/Data/news/3/6aa3f7e91ef087b5e9f647e25241745f/tourism%202012%2 0roundup%20presentation%20 (27%20mar)_3.pdf – accessed December 2013

36. Singapore Department of Statistics (2010) Census of Population, 2010: Advance Census Release, Singapore: Ministry of Trade & Industry

37. Singapore Vacation Attractions (2013) 'Top Ten Tourist Attractions: Top 10 Singapore Attractions Cover the Most Popular and Latest Attractions in Singapore': www.singapore-vacation-attractions.com/singapore-attractions.html - accessed December 2013

38. Spooky (2010) 'Haw Par Villa – The Most Disturbing Theme Park in Singapore', online at: www.odditycentral.com/pics/haw-par-villa-

the-most-disturbing-theme-park-insingapore.html - accessed May 2011

39. Teo, P and Yeoh, B.S.A (1997) 'Remaking Local Heritage for Tourism', Annals of Tourism Research v24 n1: 192-213

40. (2001) 'Theme Parks in Singapore', in Tan E.S, Yeoh, B.S.A and Wang, J (eds) Tourism Management and Policy, New Jersey: World Scientific: 233-255

41. Trip Advisor (2010-2011) 'Haw Par Villa': www. tripadvisor. com.au/Attraction_Review-g294265-d324758-Reviews-Haw_ Par_ VillaSingapore.html - accessed May 2011

42. Urry, J (1990a) The Tourist Gaze: Leisure and Travel in Contemporary Societies, London: Sage Publication, Ltd

43. (1990b) 'The 'Consumption' of Tourism', Sociology n24: 23-35

44. (1992a) 'The Tourist Gaze 'Revisited', American Behavioral Scientist v36 n2: 172- 186

45. (1992b) 'The Tourist Gaze and the 'Environment', Theory, Culture & Society v9 n3: 1-26

46. (2001) The Tourist Gaze: The New Edition, London: Sage

47. Urry, J and Larsen, J (2011) The Tourist Gaze 3.0, London: Sage

48. World Tourist Attractions (2011) 'Top Ten Tourist Attractions You Must Visit': thetouristattractions.blogspot.com.au/2012/02/top-ten-tourist-attractions-in.html - accessed December 2013

9

# GEOGRAPHY, ISLANDS AND MIGRATION IN AN ERA OF GLOBAL MOBILITY

**Russell King**

Department of Geography and Sussex Centre for Migration Research
University of Sussex United Kingdom

## ABSTRACT

This paper examines the changing role of islands in the age of globalization and in an era of enhanced and diversified mobility. There are many types of islands, many metaphors of insularity, and many types of migration, so the interactions are far from simple. The 'mobilities turn' in migration studies recognizes the diversification in motivations and time-space regimes of human migration. After brief reviews of island studies and of migration studies, and the power of geography to capture and distil the interdisciplinarity and relationality of these two study domains, the paper explores various facets of the generally intense engagement that islands have with migration. Two particular scenarios are identified for islands and migration in the global era: the heuristic role of islands as 'spatial laboratories' for the study of diverse migration processes in microcosm; and the way in which, especially in the Mediterranean and near-Atlantic regions, islands have become critical locations in the geopolitics of irregular migration routes. The case of Malta is taken to illustrate some of these new insular migration dynamics.

## KEYWORDS

geography, globalization, interdisciplinarity, island studies, islands, Malta, migration, mobility

## 1. INTRODUCTION

Geography; islands; migration: a trio of academic fields which not only define in large measure the author's own personal scholarly biography but also interlock in a triangle of creative interdisciplinarity. Islands – areas of land surrounded entirely by sea – are an obvious geographical entity, unique spaces within the world's geography. Migration – human movement across space to live 'elsewhere' – is a process which is quintessentially geographical. Geography, then, is perfectly placed to capture the interdisciplinarity and relationality of these two relatively new fields – island studies and migration studies. Simultaneously a science, social science and a humanities subject, geography is the broadest of disciplines. Indeed, perhaps it is a misnomer to call it a discipline; I would call it an 'interdiscipline'.

Islands and migration: what are the scales of the two phenomena under study? According to Baldacchino (2007: 1), some 600 million people, 10 per cent of the world's population, live on islands; and islands, including archipelagos, make up one quarter of the world's sovereign states. On the migration front, extrapolating from the latest estimates from the UN Population Division, there are 200 million international migrants in the world; that is to say, people currently residing in a country which is different from that of their birth. However, if migration within countries is included, the figure rises many-fold, but there are no global figures on internal migration.

This paper, then, is about the relations between islands and migration, seen from a geographer's broad and interdisciplinary perspective. It is in six parts. First I say something about geographers' engagement with the study of islands, which reflects the fact that most geographers regard islands as 'special' places with distinctive characteristics. At the same time, studying islands carries with it certain risks of escapism and of lack of serious commitment to the 'big picture' of world geography. Next I review in turn the two interdisciplinary fields of island studies and migration studies. Then I bring them together in a series of interactions and interfaces. Islands have nearly always had an intense involvement with migration; but the nature of this engagement, and the types and sequences of migration that result are linked to the diverse historical roles that islands have played. In the following part of the paper I develop two scenarios for island-based migration research: islands as strategic sites in the new geography of irregular migration routes, especially between Africa and Europe; and islands as spatial microcosms for the analysis of multiple migration processes and effects. The final section of the article takes Malta as an illustration of the two scenarios; Malta being an ideal case both because of the mix of different types of migration it has experienced, and because of its new role, connected with EU accession in 2004, of being on the front line of the arrival of boatmigrants from North Africa.

## 2. GEOGRAPHERS AND ISLANDS

Islands have long attracted the attention of geographers (King, 1993). Particularly notable are the early-twentieth-century writings of the French geographers Jean Brunhes (1920) and Paul Vidal de la Blache (1926), continued later by the geo-historian Fernand Braudel whose magisterial treatise on the Mediterranean speaks at length on the special role of islands in Mediterranean economic, social and cultural history (Braudel, 1972: 148-167). Brunhes (1920: 52) saw islands as "little geographical worlds" wherein "by careful study of a small unit one can discern and evaluate the relations between physical facts and human destinies". Ellen Churchill Semple, notorious for her controversial philosophy of environmental determinism, also wrote copiously about 'island people' in her book Influences of the Geographic Environment. She described islands' proneness to emigration: "A small cup soon overflows... herein lies [islands'] historical significance" (1911: 416).

These early writings, products of their time, opened up important geographical perspectives on island studies – the concept of insularity, the special characteristics of island populations (including what we would call today their identity), pioneering ecological analyses, the historical dialectic between isolation and focalism, and cycles of landscape change due to colonialism and settlement, emigration and depopulation.

Later in the last century, islands became less fashionable as objects of geographical study. As geography went robustly through its normative, model-building, theory-seeking stage, the emphasis was on grand (but often abstract) generalizations. Islands fell off the map of geographical enquiry. In fact, five problems seemed to confront geographers (and others) studying islands, and these issues are still relevant today.

- Islands equate holidays, relaxation and having a nice time, not serious scholarship. When I tell my colleagues (even the geographers) that I am going to Malta and Cyprus (the two islands where I am currently conducting some research), or to the Azores for an international conference, they smile knowingly as if to say "Oh yeah?" I insist: there is no contradiction.

- The smallness of islands denotes insignificance; there are no 'big questions' on small islands, which are regarded as "irrelevant backwaters" and "somehow unimportant" (Hay, 2006: 21; Lowenthal, 1992: 18). But geography is not about size, or the built-in assumption that large spaces are more important than small ones. No, geography is about spaces and places at different scales, about relationships between the global and the local, and about spatial relations and connectivities (including, of course, migration) in which small islands are deeply embedded and have their place in the global-local nexus (Baldacchino, 2005: 248; Ratter & Sandner, 1996: 64).

- Studying islands invokes an academic colonial gaze – the outsider, 'the expert', looking in. Baldacchino (2007: 2-3) calls this "islanding" (cf. 'othering'): "a removal of agency, cheating islanders of the possibility of defining themselves and of articulating their own concerns and interests". Thus academics, in acquiring and possessing the knowledge of the island, contribute yet another layer to islands' histories of colonialism, whose most active agent today is probably tourism1 . Grant McCall's plea that islands should be studied on their own terms (1994, 1996) suggests an escape from academic colonialism and a process of empowerment in which islanders frame the research agenda and become active partners in the research process. However, McCall's manifesto carries its own dilemmas, as we all shall see presently.

- The fallacy of regarding islands as small-scale models of the wider world. Brunhes (1920: 499) wrote that island studies were the first regional monographs; such comprehensive studies of "little worlds of humanity" pioneered the regional method in geography, leading to larger-scale studies of less definable mainland territories. Brookfield counters: "the common trap with island studies [is] of assuming that all islands are in some way a microcosm of the larger world. They are not" (1990: 31). There is, however, a middle way. Islands can be studied with rigour, both for themselves, as geographical entities (like mountains, deserts, cities… ), and as places where, under relatively controlled conditions, the interrelationships between the island environment, histories of development and underdevelopment, population and migration, can be studied in sharper focus. Herein lies the heuristic value of islands for geography.

- The danger of exceptionalism – of regarding islands as too special, too unique. This is the other side of the coin of the previous issue. Here we are faced with an essentialism of 'islandness' on the one hand, and on the other a blinkered myopia on the part of scholars about 'their' particular island which is somehow regarded as different from all others. This can be overcome both by 'horizontal' comparative studies with other islands, and by 'vertical' integration of islands within hierarchies of regional and global relations.

## 3. ISLAND STUDIES

Baldacchino (2004) has proclaimed the "coming of age" of island studies; but the essence of island studies remains both obvious and elusive. This reminds me of the old debate about the definition of geography, which often fell back on the tautalogous definition "geography is what geographers do". And "what do geographers do?" "They study the world". So it could be for islands. "What is island studies?" "It's the study of islands." What do island studies scholars study on islands?" "Everything". This implies a 'Jack-ofall-trades' approach of the combined geographer, ecologist, geologist, anthropologist, sociologist, economist and historian, enabling island studies to be "the critical, inter- and pluri-disciplinary study of islands on their own terms" (Baldacchino, 2007: 16).

Let me now head off at a bit of a tangent. Island studies seems to be a male domain (hence my use of 'Jack' above was deliberate), in contrast to migration studies which has become feminized, as we shall see below. Despite (or indeed because of) the fact that islands themselves, with their images of tropical climates and swaying palms, are often perceived as 'she'/female, islands are objects predominantly of the male gaze, to be studied, objectified, collected and possessed. I know island aficionados (males, of course) whose islophilia involves visiting as many islands as possible, ticking them off the list, and collecting the T-shirt to prove they have 'been there'. Baldacchino (2007: 3-4) comments on the desire of the (almost inevitably male) visitor to circum-navigate the island, climb triumphantly to its highest point to 'take it all in', even to sing and shout. According to Hay (2006: 27) islands connect with metaphors of western masculinity, as well as literary classics ranging from Robinson Crusoe (Defoe 1719/1980) to D.H. Lawrence's The Man who Loved Islands (1926/1982)2 .

More prosaically, McCall (1994, 1996) has proposed the term nissology for "the scientific study of islands on their own terms". Along with Baldacchino (2008a), I detect an inherent contradiction in this definition. McCall favours the islanders' own emic perspective over the etic gaze of the outsider, but two problems arise. First, "scientific study" implies a set of orthodox research strategies which may override the islanders' own view. Second, the division between islanders and non-islanders, insiders and outsiders, is not so clear-cut. Given islands' long histories of migration, both in and out, who is an islander? What, therefore, is meant by "on their own terms"? Such a construction implies homogeneity and agreement within island communities, and overlooks divisions based on class, ethnicity or migration history.

Another approach to island studies is through the notion of 'islandness' (Baldacchino, 2004: 272-274), the key characteristics of which are usually listed as smallness, a strong sense of local or island culture and identity (typically implying some tension with a larger 'mainland'), peripherality, and a sense of being bounded by water. But parts of this framing of islands project a too-negative image, as Baldacchino himself acknowledges. The term 'insularity' carries problematic connotations of isolation and inward-lookingness, whereas many islanders are very outward-looking, by their very nature as islanders. Peripherality is both relative and subject to change, as well as being multi-dimensional.

Taking Europe as an example, the economic geography of core and periphery under one system (the Fordist model, for instance), may become partially inverted under another economic model, based on tourism, light industry and financial services. The 'Celtic Tiger' economy of Ireland is a case in point, propelling the country within a decade from a relatively poor status (by European standards) to one of the continent's most prosperous and dynamic member-states. Or consider Ibiza: transformed from a rural, semi-subsistence economy to the youth tourism clubbing capital of Europe.

A third angle on island studies which is quite prominent, especially in the geographical literature, is the 'spatial laboratories' approach, the dangers and possibilities of which I noted earlier, including the fallacy of regarding islands as miniature replicas of the mainland or as closed systems. Nevertheless islands have long been productively utilized as a semi-laboratory setting for the study of ecological, epidemiological and social processes, including migration (Connell & King, 1999; King, 1999).

As regards migration, two further approaches to islandness are especially relevant – economics and demographics. There is now an extensive literature on the economic and developmental problems of small islands, some of which overlaps with economic analyses of small states (see, inter alia, Baldacchino & Greenwood, 1998; Baldacchino & Milne, 2000; Bertram & Poirine, 2007; Biagini & Hoyle, 1999; Briguglio, 1995; Connell, 1988; Dommen & Hein, 1985; Hintjens & Newitt, 1992; Lockhart et al., 1993). Undoubtedly, smallness and insularity bequeath a range of economic handicaps3 : the small size of the local market for goods and services; the limited range of island products due to the narrow resource base; high transport costs for both exports and imports; vulnerability to world market prices; risk of crop failure due to exposure to natural disasters such as drought, hurricane or diseases; and the limited or 'truncated' local labour market with few opportunities for highly educated or ambitious young people. Emigration often becomes the response to these economic vulnerabilities; labour becomes the 'monocrop' to be exported to the global market, as Schneider and Schneider (1976) showed in the context of western Sicily, where workers replaced wheat as the major export product. Especially for smaller islands, the end-result is often the MIRAB economic model, whereby migration, remittances, aid and bureaucracy are the key parameters (Bertram & Watters, 1985; for updates see Bertram, 2006; Bertram & Poirine, 2007).

But this is not the whole story. Islanders' supposed condition of poverty and economic vulnerability is countered by the fact that island states' GDP per capita is above the world average, and small island states are disproportionately numerous amongst higher and upper-middle income countries (Read, 2004). The economic riddle is solved by islands' skills in identifying high-income-earning niche products and services in the global economy, trading on the economic nimbleness which is perhaps easier to exercise from a small-scale jurisdiction (Bernard & Poirine, 2007). Islands thus exemplify post-industrial spaces where flexible specialization is the watchword and where a higher level of connectivity to the global system can be capitalized upon. Tourism, including niche tourism, and financial services, especially offshore banking, seem to be the key sectors for many islands' prosperity, although such activities are, in turn, not without their risks and limits. Tourism and banking are themselves subject to market volatility, and not every island is equipped climatically for tourism, or knowledge-wise for financial industries. Tourism in particular brings its own problems of seasonality, disruption of island lifestyles, external ownership and environmental pressure – there is extensive literature on these issues (e.g. Apostolopolous & Gayle, 2002; Briguglio et al., 1996a, 1996b; Gössling, 2003; Gössling & Wall, 2007; Lockhart & Drakakis-Smith, 1997).

The final analytical field within island studies – one which has a close bearing on migration – is the socio-demographic perspective. Many small islands evolve relatively homogenous and intimate societies with a strong sense of common identity. Everyone knows everyone else, at least as an acquaintance or through common friends, and there is an equally close knowledge of the local environment – every house, every field, every track, every beach is known and recognized 4 . Population growth may stretch this equilibrium in various ways, leading ultimately to out-migration. Pressure on the land, meantime, can provide the stimulus to construct elaborate terrace systems, seen in various parts of the world: in Java for instance where the progressive increase in the population led to a parallel intensification of the agricultural ecosystem with spectacular rice terraces and minutely managed irrigation systems, a process called by Clifford Geertz (1963) agricultural involution. Elaborate terraces for vines and other crops are also found throughout the Mediterranean and Macaronesian archipelagos; perhaps most spectacularly on Madeira. Eventually, the demographic momentum may overshoot, projecting the population beyond the island's carrying capacity at the level of technology and economic resources that the island possesses at the time, and emigration results. This leads to new cycles of demographic, social, economic and landscape change which have been copiously analysed in the literature on island migrations (for overviews see Connell, 2007; Connell & King, 1999; King 1999). Return migration may cushion the long-term demographic and economic decline somewhat, but much depends on the age of the returnees and the financial and human capital they bring back.

## 4. MIGRATION STUDIES

One of the justifications for the 'coming of age' of island studies as a recognized academic field is the existence now of two MA programmes (at the University of Prince Edward Island and the University of Malta) and three dedicated journals, the peer-reviewed Island Studies Journal, which started in 2006, Shima: The International Journal of Research into Island Cultures, started in 2007, and the longer-established (since 1992) but less academic Insula, published with the support of UNESCO (Robertson, 2007: 576-577).

On these criteria, 'migration studies' is an older and more broadly established interdisciplinary field. The first MA in Migration Studies started at the University of Sussex twelve years ago, and there are many such programmes now, especially in Britain and Europe, including some specialized masters' such as Oxford's MSc in Refugee Studies. The three leading journals in the field – International Migration, International Migration Review and the Journal of Ethnic and Migration Studies – have been running since 1962, 1966 and 1974 respectively. New journals are founded every year, it seems, one of the latest and most interesting being Mobilities, launched in 20065 .

The literature on migration studies is truly vast and growing exponentially. Following my approach in the previous section, I signpost some key themes and perspectives.

Migration is usually defined as the movement of people from one place or country to another, for a significant period of time such as more than one year. It is, thus, both a spatial event and a temporal one; but also to some extent an artifice of the spatio-temporal thresholds that have to be crossed for migration to be said to have taken place. For the statistical measurement of migration, boundaries have to be crossed (international frontiers, or internal administrative boundaries such as provinces or municipalities), and certain time(s) elapsed before the move is 'registered'. Hence the measurement of migration is far from uniform across the world. Thresholds are arbitrary and open to different interpretations in different places. In the United States, for instance, a move from one street to another within a city, or from one rural area to another within a county, might not be conventionally regarded (or statistically recorded) as migration; it might instead be called 'residential relocation'. Time, too, is arbitrary and situation-dependent when it comes to migration. The one-year criterion noted above – which is commonly used by migration scholars and national and international statistical agencies – leaves out seasonal migrations whose movement regimes are more short-term but no less significant, both to the individuals concerned and in terms of their contribution to the overall map of global human mobility.

Although migration has been a constant feature of humankind throughout history (King, 1996, 2007), the academic study of migration really only started with the industrial age (Boyle et al., 1998: 59). It was the geographer Edward Ravenstein who made the foundational statement about migration studies in his famous 'laws of migration' (1885, 1889), still widely referred to today. Ravenstein's eleven laws – which are really empirical generalizations rather than laws sensu strictu – showed migration to be essentially economically driven, both in terms of individuals' motivation and with regard to the relative levels of development of sending areas (poor, rural, with limited opportunities) and the receiving contexts (urban, industrial, commercial, and with more opportunities). One of his laws – that females are more migratory than males, particularly over shorter distances – passed almost unnoticed at the time but has been shown to have special relevance in recent years. I refer here to the feminization of migration, by which I mean three things. First, females make up an increasing share of total migrants. Nowadays, females comprise almost exactly half (49%) of all international migrants: although, to be fair, and this was Ravenstein's point, females have always been prominent in migration streams (it was simply that they were 'ignored' by theories and narratives which constructed migration as a 'male' process). Secondly, women have increased agency in migration; that is to say they are increasingly migrating independently of men and often as leaders of migration chains. This is in contrast to the past where they were more likely to be (constructed as) 'followers' or 'accompaniers' of the 'pioneering' male migrants. And thirdly, the study of migration has become feminized. Whilst it is true that the older generation of migration scholars (like island studies scholars) remains resolutely male-majority, amongst younger researchers there is a distinctive majority of females. In contrast, then, to the masculinist gaze of island studies, the study of migration is increasingly, and predominantly, carried out by (young) women

Of course, there is much more that could be said about migration; far more than can even be hinted at here. Let me just make a few key points. Following Castles & Miller (2003: 21), I make a distinction between the act of migration, and migration's consequences; or between causes and effects. This has special relevance to islands where the causes and consequences of migration may be clearly separated in space and time. Causes embody several fundamental questions. Why do people migrate? Who migrates? And where do they move to? The effects fall into two groups: impacts of migration on the places/islands 'left behind' (depopulation, demographic distortion, brain drain, loss of labour etc.); and the formation of migrant communities in the places of destination (such as the preservation of 'island identity', integration into the 'host society', and possible return 'home'). However, the separation between causes and consequences is not absolute. Consequences can become causes, for example when depopulation (an effect) causes further outmigration because of the declining market base for economic activities, or declining morale. Likewise, the return to the island of 'successful' migrants can act as a stimulus to further migration through the 'demonstration effect' 'Cause' and 'effect' is only one of a number of binaries which dissect the study of migration. In practice, these dichotomies are often blurred, as the following examples show (King, 2002).

Forced vs. voluntary is a common division in migration studies, used particularly to address the distinction between refugees and so-called 'economic migrants'. In recent years the latter term has come to have a negative connotation in the media and in political rhetoric (especially in the UK), and has become infused with implications of 'illegality' and 'bogus' asylum claims (Kaye, 2001). Of course, there are migrations which are easily classifiable either side of this binary. The forced evacuation of the population of Diego García in the Indian Ocean to make way for a US military base is an obvious example (Vine, 2004). Equally clear-cut as an instance of voluntary migration is the 'retirement migration' of older British and German citizens to islands such as Majorca and Tenerife (Casado-Díaz et al., 2004). But a great deal of migration that takes place in the world lies between these two extremes and blurs the forced/voluntary divide. The history of migration from many Caribbean islands, especially in the early post-war decades, is a good illustration. Migrants were not coerced to leave as such, but were, in effect, driven to emigrate by powerful economic imperatives. For the poor and unemployed, emigration was the only 'way out' of destitution and the only route to achieve a measure of well-being

A further important dichotomy is that between internal and international migration. As Skeldon (2006: 16-17) points out, these amount to "two migration traditions" based on different datasets, theoretical assumptions and analytical methods, producing different literatures and linked to different policy arenas. Yet this dichotomy is, in practice, blurred – for instance when the same migrant moves both internally and then internationally (or vice-versa), or when boundaries change, so that former internal migrants become international migrants (as in Yugoslavia before and after the creation of new independent states in the 1990s).

Introducing the island dimension further complicates, but also illuminates, this division of scale and type of migration. Imagine an archipelago made up of larger and smaller islands, not an independent state but part of a country whose main territory is a 'mainland': the Azores is a good example. The following scale hierarchy of migrations may be observed:

- intra-island migration – especially on a large island like São Miguel, there will be local-scale migrations from remote villages or hamlets to the capital Ponta Delgada and other main towns;
- inter-island migration – for instance from smaller islands like Santa Maria or Graciosa to larger ones such as São Miguel or Terceira;
- migration to the mainland – this is still internal migration within the same country (Portugal) but may involve considerable distances and 'dislocations';
- migration to another country – this could either be on the same continental mainland as the country to which the archipelago belongs (i.e. another European country such as France – actually this is not an important destination for Azoreans although it has been for Portuguese emigrants), or another country in another continent (in the Azorean case, most emigration has been to the United States and Canada, plus some other countries such as Brazil and Bermuda).

Once again, we need to acknowledge that these migration trajectories are not necessarily mutually exclusive alternatives but can be combined in an individual's migratory biography in various sequences, perhaps interleaved by return migration(s), or they can be experienced by different members of the same family. Nor is the above list an exhaustive typology: reverse migration flows, such as return from abroad, or the movement into rural island spaces of dissatisfied urban refugees, are also part of the migration dynamics.

Finally, to bring this briefest of overviews of migration studies up to date, two new paradigms have coursed through the field over the past 15-20 years. Both reflect the broader phenomenon of globalization and its effect on mobility and identity. The first is the transnationalist approach (see Basch et al., 1994 for the definitive study), which has, to some extent, rejuvenated the older tradition of diaspora studies (Vertovec & Cohen, 1999). Transnational migrants are immigrants who engage in regular and sustained activities linking them back to their homeland. It is not enough to be an international migrant to be transnational: the key criterion is that transnational activities (sending remittances, making return visits, supporting hometown associations, trading goods, close communication with kin back home etc.) are a central part of a migrant's life. With their usually strong islander identity, it is often the case that island-origin emigrants have strong transnational identities, although translocal might be a better descriptor. As the title of one of my co-edited books indicates, island migrants come from 'small worlds' yet lead 'global lives' (King & Connell, 1999). Their global lives are rooted in the small place that is 'their' island, whose community they still see themselves part of, even if their returns are sporadic or negligible (Felgentreff, 1999). As a popular Azorean saying goes, "Azoreans always carry their island on their backs" (Williams & Fonseca, 1999: 67). And it is no accident that some of the pioneering studies of transnational migrants in the US are based on islander migrants, notably Dominicans and Haitians (see Glick-Schiller & Fouron, 1998; Guarnizo, 1994, 1997; Itzigsohn et al., 1999).

The second, and more recent, new paradigm in migration studies is the 'mobilities turn' (Sheller & Urry, 2006; Urry, 2000) which not only stresses the defining role of mobility in postmodern societies but also frames migration as one of a continuum of mobility types, including travel, tourism, circulation, sequential migration, and virtual mobilities. Urry (2000) and Beck & Beck Gernsheim (2002) argue that what characterizes global society, and especially the more advanced societies of the world today, is not the old 'fixed' categories of class, residence, career-based employment, family and kinship, but rather individualization and mobility. People increasingly story their lives not so much as who they are, what they do and where they live, but recount where they have been and where they want to go. Thus, alongside the recognized 'forms of capital' – financial, human, social, cultural (Bourdieu, 1986) – we can now consider mobility capital. Islands feature prominently in these new 'mobilityscapes', both as origins (islanders who want to leave, 'see the world' and return) but more especially as destinations for tourism, travel, escapism, 'dropping out', 'moving in and moving on', and so on.

Like other forms of capital, the distribution of mobility capital is highly uneven. How people deploy their mobility capital depends on whether they are rich or poor, and on where in the world they were born and hence what 'mobility rights' they have. It is an irony that, in this supposedly globalized world of free movement (of goods, capital, ideas, and so on), one of the key factors of production – labour, people – is subject to draconian regimes of migration control. Or rather, certain categories of people (the wealthy, those from 'the West', people of certain ethnicities) are free to migrate, whereas others are not (the poor, the uneducated, people of colour, from the less developed world). This polarization in access to mobility seems to be getting worse, so that there is an increasing gap between the aspiration to migrate and the ability to do so (Carling, 2002). Castles & Miller (2003) write that we live in "the age of migration", but this need to be qualified: an age of migration for some, but not for others.

## 5. MIGRATION AND ISLANDS

I rack my brains and I cannot recall any definitive study of migration from an island where the island dimension is central to the study. (Of course, there have been very many studies of migration from islands, but none are truly canonical). If we look at accounts of the history of migration, we find them dominated by the so-called 'great migrations' – of slavery, indenture, colonial settlement, labour migration (King, 1996). Islands have been involved in all these but are not a key part of the grand narrative

Yet there is no doubt that islands have nearly always had an intense engagement with migration – settled by migration in the first place and often deeply affected by migration thereafter. The rates and indices of migration on islands – especially as regards emigration/outmigration – are certainly higher than 'mainland' countries or regions (even if no-one has made the exact calculations). The intensity, types and chronological sequence of migration depend on the historical roles that islands have played. In a previous essay (King, 1999) I drew a simple contrast between two historical roles: islands as marginal spaces, and as nodal places. Nodal islands tend to attract and exchange population, leading to the creation of cosmopolitan, hybrid and stratified societies, often with an 'open' mentality towards the outside world. Marginal islands tend eventually to become areas of outmigration and depopulation, leading to the MIRAB syndrome described earlier.

But marginal vs. nodal is a too-simplistic dichotomy of islands in the current global scenario. Bertram & Poirine (2007) show that the smaller and more isolated the island, the greater the need to be open to the world market and to specialize in a narrow set of revenue-generating activities. For successful small-island economies, the keywords are specialization and flexibility; especially the ability to make rapid switches to new and profitable specialized economic niches. Bertram & Poirine (2007: 333-335) point to the rapid shift of the Cayman Islands from a classic MIRAB economy to a financial services and tax haven powerhouse within a decade. Hence the Cayman Islands have virtually no emigration and, instead, attract in-migrants and visitors from near and far (Connell, 1994). These authors demonstrate that the economic vulnerability index (a measure of the narrow range of economic activities), based on data from 69 island economies, is positively not negatively correlated to per capita income.

Warrington & Milne (2007) proffer a more sophisticated typology of islands, based on governance and historical relationship to the 'world system'. Below, I follow their sevenfold typology, making connections to different histories of migration. These types are by no means mutually exclusive, and certain islands can fulfil different roles at different points in their history, or exhibit a combination of roles simultaneously.

First come islands of civilization, of which, in modern times, there are just two, Britain and Japan, both large archipelagic countries. Through trade and conquest, these islands have generated global economic and cultural power, and an insular autonomy and hegemony that are the very antithesis of the vulnerability so often regarded as typical of islands. Linked to their imperial political and economic projects, these islands' migrations are those of colonial and trading settlement. Later, especially in the case of the UK, labour inmigration from the (former) colonies and other poor countries becomes characteristic, driven by the demands for labour that a highly prosperous capitalist society generates.

Next we have islands of settlement: these are settler colonies which are recipients of the colonial migrants referred to immediately above. For Britain, they were, above all, the 'white dominion' islands of New Zealand and the island-continent of Australia. Iceland is another example, from another imperial realm. Their role has been one of thinly populated territories capable of absorbing surplus population from the colonial homeland, mainly through voluntary and often sponsored and subsidized migration

Third are plantation islands, also a product of the colonial era, but this time as more subordinate elements, supplying a narrow range of commodities to the imperial homeland. Typical plantation products were sugar, cotton, coffee, tea, spices and valuable minerals. This hyper-specialization persists in the form of other 'monocrops' – hydrocarbons, tourism and (arguably) offshore finance. Economic vulnerability remains a structural characteristic of plantation islands, not just because of narrow specialization and dependence on world market prices for primary products (which have generally been falling), but also because of heavy concentration of ownership of the means of production either in the hands of a local plantocracy or in multinational corporations (Warrington & Milne, 2007: 411). Migration-wise, such islands are characterized by a history of involuntary immigration of slaves and indentured labour, which has bequeathed a racially structured unequal society. When the economy collapses, mass emigration results; labour becomes the monocrop to be exported. Many of the island states and dependent territories of the Caribbean (and also some in the Indian and Pacific Oceans) conform to this regime of political economy and migration.

Islands as fiefs share some of the historical, economic and migration characteristics of plantation islands. Fief islands are the polar opposite to islands of civilization. They are defined by extreme peripherality and vulnerability compounded by a history of rapacious exploitation by local and external elites. Sicily and Haiti are archetypal fief islands, afflicted by a set of structures which are inimical to development – mafia organizations, extreme traditionalism, violence, corruption and illegitimate exercise of power. Irish history before 1921 is also often defined in terms of being a British protestant fiefdom. Wholesale emigration, often under traumatic circumstances (cf. the Irish famine of the mid-1840s), is the accompanying consequence as the only 'way out'.

Next comes the island as fortress; a site for protecting and projecting power. Malta is the exemplar, but there are other cases such as St. Lucia in the eighteenth century, Singapore during the interwar years, and

Hong Kong prior to its transfer to China. Colonial and military power are the key to the raison d'être of the fortress island, based on a fortuitous combination of strategic location, maritime and air travel technology (which of course can change), and trade and military routes. According to Warrington & Milne, the pattern of governance of fortress islands is replete with paradoxes: "economic affluence and vulnerability; competent, benign administration but fractious politics; a cockpit of imperial politics, but a weak sense of domestic interests" (2007: 407). The build-up of the fortress economy usually involves some military in-migration, but then mass emigration (as in post-war Malta) follows when the fortress function wanes.

Entrepôt islands exhibit a clear nodal vocation. Locationally entrepôts benefit from nearness to incoming trade routes, functioning as a semi-detached entry-point for adjacent mainlands; but internal sources of entrepôt success derive from investment finance and entrepreneurial flair, and a regulatory regime that facilitates market transactions by minimizing costs and risks. Warrington & Milne (2007: 413) quote Singapore, the Channel Islands, the Isle of Man and Mauritius as examples of this island political economy. For others – Barbados, Malta, Ireland – the entrepôt function may take over from other political economies – respectively plantation, fortress or fief. The entrepôt trades on commerce and investment rather on large-scale migration, but a combination of emigration and immigration, especially of wealthy merchants, entrepreneurs and tax exiles, is to be expected.

Finally there are islands of refuge, which are conceived as bulwarks against dominant mainland powers. Cuba (against the USA) and Taiwan (against mainland China) are the classic examples, defining themselves ideologically against the alter ego: socialism against capitalism, democracy against autocracy. They are characterized by migrations of refuge, both in and out.

## 6. ISLANDS, MIGRATION AND GLOBALIZATION: TWO SCENARIOS

Set against the dual background of the current era as both the age of globalization and the age of migration, and with my main geographical reference being the Mediterranean Sea and the adjacent eastern Atlantic, I pull out two research scenarios for islands and migration.

The first focuses on the new and key functions of islands as strategic locations in the geopolitics of irregular migration, especially as regards the movement of 'clandestine' migrants between Africa and Europe. Many migrants seeking to enter Europe are engaged in long and tortuous journeys lasting several months or even years: their origins lie far to the south of the African shore and the target destinations may be countries beyond the southern European coasts and islands. Indeed, as 'transit migrants' involved in 'fragmented journeys', their country of destination may not be determined when they leave home; it may change en route according to opportunities that present themselves; or, indeed, they may never get there but be blocked in some intermediate location (Collyer, 2007: 668). Poor migrants (and not-so-poor migrants, given that irregular migration involves substantial costs of travel and smugglers' fees) are increasingly driven by 'push pressures' to depart from their home countries which are often in political or economic chaos; yet they are confronted by global/European regimes of migration control and border securitization which make entry into 'Fortress Europe' ever more difficult and dangerous. As new routes of 'least resistance' are increasingly sought out by migrants and smugglers, certain islands offer themselves as strategically-located staging-posts and entry-points, especially those which are part of the European Union and yet are close to the 'launching pads' of migration in Morocco, Tunisia, Libya etc. In this new strategic geography of irregular migration, Lampedusa (the tiny outpost island of Sicily and Italy), Malta, and the eastern Canary Islands (especially Fuerteventura) are the most attractive stepping-stones into Europe.

The attractiveness, but also dangers, of islands as key sites in the journeys of irregular migrants are all-too evident when we think of the precise spatialities of migration routes and migration control. Controlling a maritime border is fundamentally different and more difficult than a land border because it requires surveillance of an area (the sea) rather than just a line (Carling, 2007a: 324). Compared to land borders, with their checkpoints at road crossings and airports, and the possibility of erecting physical borders such as fences and walls, sea borders are more open and access can be stealthier, under cover of darkness. 'Search and rescue' zones make it possible, even obligatory, for migrants in overloaded boats who are in mortal danger to be picked up and 'taken in', with the chance that they will not be repatriated. The dangers derive from storms and currents overturning the flimsy boats or driving them off-course, engine failure, poor navigational skills, shortage or absence of life-jackets, inability to swim, as well as the unscrupulous behaviour of smugglers and pilots who may abandon migrants too far from the shore to the peril of those who cannot swim.

All this has been amply documented, not so much in academic studies (but see Carling, 2007a), but in a plethora of newspaper and NGO reports7

Carling (2007a, 2007b) and Collyer (2007) have recently documented the evolving geography and technology of migration control in the Mediterranean and near-Atlantic seas. The imposition by the Spanish authorities, with EU backing, of the SIVE surveillance system (Sistema Integrado de Vigilancia Exterior) across the Gibraltar Strait in 2002 made the short-sea crossing between Morocco and Spain, which for years had been crossed by thousands of Moroccan and other migrants in flimsy pateras (makeshift wooden boats), increasingly less feasible. This had the effect of deflecting the main routes for transit migration between Africa and Europe: eastwards to the stretch of Moroccan coast between the Spanish enclaves of Ceuta and Melilla, with migrants aiming at the coast of Granada and Almería; and south-westwards down the Atlantic coast of Morocco, aiming at Cadiz. Application of SIVE to the entire Andalusia coast then forced migrants even further south and east. Routes developed in the early-mid 2000s from southern Morocco, Western Sahara and Mauritania to the Canaries, initially aimed at Fuerteventura, the closest to the African coast, but then, after a partial introduction of SIVE, to Lanzarote, Gran Canaria and Tenerife. As surveillance intensified off these coasts, longer-distance routes have very recently sprung up, involving the Cape Verde islands far to the south. Meanwhile, far to the east, a new maritime bridgehead opened up for irregular migrants transiting Libya towards Sicily and Malta, the latter a new EU entrant in the 2004 enlargement. The case of Malta will be dealt with in the final section of the paper.

Before that, I turn to the second island research scenario. This harks back to the remarks made earlier about the heuristic value of islands as small-scale laboratories for the analysis of societal processes and interactions, but it also has a contrastive connection with the first scenario just outlined. The first scenario was mainly about poor migrants travelling without papers, searching out viable routes to access rich European countries to find work and improve their lives and those of their family members back home. We saw how some islands have acquired great locational significance in this new migration geography. However, for individuals at the other end of the global wealth spectrum, islands become 'playground spaces', to visit as tourists or on yachts or cruises, to buy second homes (or third or fourth homes...), or even to buy the whole island! The invasion of foreign capital to buy island property sets in motion a process of island gentrification which has been imaginatively documented for the

Swedish islands around Gothenburg and Stockholm by Clark et al. (2007), and also for Capri (Mazetti, 2007). According to local estate agents, the Swedish market for property on islands has become 'hysterical', such is the demand (Clark et al., 2007: 483). Meanwhile, on the isle of Capri, whose origins as a holiday playground can be traced back to Roman times, building land costs between 6,000 and 10,000 euros per square metre, depending on the site and the view. These are realestate values comparable to many central-city locations, so that Capri has in effect become a luxury residential district of metropolitan Naples, to which it is connected by fast and frequent hydrofoil services (Mazetti, 2007: 612).

In this role, and very different from the vision of desperate migrants, bedraggled on the shores of Lampedusa or Fuerteventura, islands offer an altogether more romantic image which builds on 'the island' as a constant visual and metaphorical trope of tourism and relaxation. Islands embody unique small spaces where peace and quiet, environmental purity and pristine landscapes and seascapes are especially targeted by rich people, retirees and dreamers who have the wherewithal to relocate there, at least for part of their lives.

All these migration types – rich and poor migrants, short-stayers and long-stayers, tourists and retirees etc. – can be located together on many islands such as Tenerife, Gran Canaria, Majorca, Malta or Cyprus. To a greater or lesser extent, they all interact both with the host society and with each other. To be more categorical, we can recognize the following actor groups in this scenario:

- island-born non-migrants – who are nevertheless profoundly affected by the migration of other islanders, as well as the presence of immigrants;
- internal migrants within an archipelago – these may not be immediately recognizable but may partly contribute to define the boundaries of 'belonging';
- emigrants – who may make regular return visits and continue to see themselves as part of the island community;
- returned migrants – who may have returned temporarily or for good;

- tourists and visitors – ranging from short-stay tourists and cruise-ship passengers to those who have second homes where they may stay for several months a year;
- immigrants from wealthy countries, who may be retirees; and
- immigrants from poor countries – both documented and irregular, including asylumseekers.

The small-island context allows close study of multiple migration processes and their interactions; this is more easily visualized within the insular microcosm where the migration actor-groups listed above live in close geographical (but not necessarily social) proximity to each other.

Several interesting research questions and hypotheses arise. Does the small-scale island society offer a welcoming environment where migrants (of various types) can easily settle, supported by personalized relationships which are easier to develop in a local, informal setting? Or are islands unwelcoming places to outsiders who are seen as potential threats to established social orders and networks (see Baldacchino, 2008b)? Or perhaps some immigrant groups are welcomed, others tolerated, and others rejected? How is the key triangular relationship between islanders, wealthy foreign settlers and poor labour immigrants played out? Are returnees 'intermediaries' between 'natives' and immigrants, drawing on their own experiences of migration to broker understanding between the various groups?

As yet, there are very few island-based studies which provide answers to these questions. One exception is Salvà-Tomàs (2002) who has examined the relationship between tourism (largely from Germany and the UK) and labour migration (from Africa and South America) in the Balearic Islands, although his is largely a descriptive economic and demographic analysis, not one based on primary survey or ethnographic data. More theoretically innovative is Bianchi's (2000) study of the 'tourist-worker nexus' in the Canaries. Bianchi problematizes the distinction between tourists and tourist-industry workers by exploring the hybrid category of transient resort workers in the context of emergent patterns of postindustrial mobility in which the boundaries between work, leisure and migration are blurred. "Migrant tourist-workers are thus a distinct [group who] encompass aspects of both travel- and work-related mobilities in so far as they lodge their sense of identity in the pursuit of alternative, transient lifestyles, in which resort work and recreation are integral components" (2000:

122). Although this sociological category is not unique to 'holiday islands', the sense of escapism associated with island tourist destinations draws many such tourist-workers to island hot-spots such as San Antonio (Ibiza), Mykonos or Agia Napa (Cyprus), where seasonal and casual jobs as tourist reps, nightclub staff, bar workers, lifeguards etc. are fairly abundant.

## 7. THE EXAMPLE OF MALTA

Malta epitomizes both the island migration scenarios referred to above: the mixing of different migration actor-groups and types of migration within a small insular space and society; and the role of islands in the new spatial dynamics of clandestine migration into the EU. Of course, no island is 'typical' and Malta's unique geo-historical features must be recognized, notably its extremely high population (for its size) and its harbour of special historical and military importance.

The notion of Malta as 'migration's microcosm' (Anon, 1964) is built on a long and varied historical sequence of migrations affecting the island. Malta's function as an island fortress at the crossroads of the Mediterranean – part of a trio of British imperial defence-posts in the Basin along with Gibraltar and Cyprus – gave the island a rather particular developmental and demographic history. The externalized economic base, on an island with few natural resources and an already-dense population, led to a rate of population increase which ultimately proved unsustainable, especially when the 'fortress economy' started to shrink after Britain's global rundown of its military bases following the Suez debacle in 1956. However, long before the post-war emigration boom, Malta had a tradition of exporting its surplus population, especially within the Mediterranean Basin via its commercial links to prominent port-cities such as Algiers, Tunis, Tripoli (Libya), Marseilles and Constantinople (Istanbul)8 . Price (1954: 115) estimated that, by 1865, there were 40,000 Maltese living abroad, equivalent to 22% of the population.

This early migration failed to curtail a population growth rate which remained remarkably high for a European country (albeit a strongly Catholic one ideologically opposed to birth control) until well into the post-war era. According to the 1948 Census, one quarter of completed families had ten or more children, a proportion which dropped to one-fifth in the early 1960s, when average family size stood at six children per completed family. Meanwhile the death and infant mortality rates dropped vertiginously, exacerbating population growth. During the late 1950s and early 1960s, births averaged 8,600 per year, deaths only 2,820, yielding a net annual increment to the population of nearly 6,000, or 2% of the population, very high for a country of Malta's level of development and standard of living (King, 1979: 110-111).

The push factors for Malta's 'great exodus' during the three post-war decades were both demographic (high natural increase) and economic (contraction of the fortress economy), the former stressed by Jones (1973), the latter by Delia (1982). During 1946-74 there was an exodus of 138,000, equivalent to 44% of the Maltese population. Four destination countries absorbed these emigrants: Australia (58%), the United Kingdom (22%), Canada (13%) and the United States (7%). During the peak years of emigration – the 1950s and early 1960s – the rate approached one in 30 or 3% per year, the highest emigration rate in the world, according to Jones (1973).

As emigration tailed off during the 1970s, two new migrations appeared: a fairly substantial return migration of Maltese labour migrants (King, 1979; King & Strachan, 1980), and the immigration of so-called 'sixpenny settlers'. The latter were retired (often early-retired) British nationals, many with a military background, who were attracted to Malta by prior connections (such as wartime service there), its low tax rate (six old pence in the pound, or just 2.5 per cent), the warm and sunny climate, and its 'British' colonial flavour (even after independence). This immigration of British retirees continued into subsequent decades; in the mid-1990s the British High Commission estimated the size of the British population in Malta at 5,000 (Warnes & Patterson, 1998: 121).

There is, however, a more dramatic and recent story of immigration into Malta. This is associated with the country's entry into the EU in 2004, as well as with the above-noted rerouting of African irregular migrants via Libya as a major new transit country9 . This new immigration trend was already evident in 2002, when Malta's 'European' vocation, after years of debate and vacillation, had become clear, albeit with residual overtones of ambivalence (Baldacchino, 2002; Mitchell, 2002). During that year, 21 boats deposited 1,686 irregular migrants and asylum-seekers on the shores of Malta – a rate of arrival which has remained more or less constant, at around 1,700-2,000 per year, ever since. The main nationalities are Egyptians, Sudanese, Ethiopians, Eritreans and Somalis.

The sea journey to Malta, for which migrants pay agents around US$1,000-1,200 (figures for 2003-05: Hamood, 2008: 30), follows an equally perilous and much longer (but cheaper, US$60-180) overland journey on open trucks through the desert, crossing into Libya via unguarded border points. Ironically, most of the migrants arriving in Malta do not want to be there; their target is Italy and the open European space of the Schengen area. Their arrival in Malta is by mistake: they think, or are told by the pilots, that Malta is Sicily; or they are picked up if in danger by patrol-boats in Malta's extensive search and rescue zone – a vast elongated rectangle which stretches from just off the coast of Tunisia nearly to Crete and which is hugely out of proportion to Malta's size compared to the adjoining countries of Tunisia, Libya, Italy, and Greece (see the map in Texiere (2006: 42). The migrants are then 'trapped' in Malta because the rules of the Dublin convention state that asylum claims must be processed in the country of first arrival, with no onward mobility allowed for claimants, and because Malta (unlike Italy) has no return agreement with Libya (Hamood, 2008: 32-33).

Immigration has rapidly proved to be a challenging and hotly debated issue in Malta, stirring up reactions which can certainly be labelled as xenophobic. Neil Falzon, United Nations High Commissioner for Refugees representative in Malta, has stated:

"There's an ugly xenophobia developing here and the government carries some responsibility for that. It is selling the idea that Malta can't cope. The truth is it has to... The government should be leading the process of integrating [the immigrants] with jobs, education and homes instead of taking part in this national hysteria" (quoted in Jones, 2008).

Although the debate on immigration in Malta is often expressed very simplistically, the underlying issues are complex. Undoubtedly there is a wider trope of 'invasion' which is linked both to claims of smallness, islandness and high population density; and to a 'fortress history' of fear of invasion, which also became apparent in debates surrounding EU accession in 2004 (Baldacchino, 2008b: 14). Immigration discourse is also refracted through historical templates of 'race' and 'nation' in which certain immigrants and visitors (especially tourists, Europeans, Christians) are accepted, whilst others (Africans, Arabs, Muslims) are not. Yet seeing immigration as a threat to Maltese national identity is inherently problematic because of the very nature of the concept of identity – identities, pace Stuart Hall (1996), are always contested, insecure, manufactured, processual, relational ... – and because of the consequent difficulty of pinning down exactly what a Maltese national identity – if it exists – consists of (Abela, 2006; Baldacchino, 2002). Bearing in mind these inherent complexities surrounding the nature of Maltese national and cultural identity, I now proceed to try to analyse the discursive dynamics surrounding immigration more systematically.

The recent influx of irregular migrants challenges Maltese national identity and migration policy in three ways. First, there is the perceived 'intrusion' of migrants of different cultures, religions and skin colours, and mostly single young males, into a society which has generally regarded itself as homogenous, family-centred, staunchly Roman Catholic and socially conservative (yet also, at the same time, progressive towards its own disadvantaged members). But this homogeneity is tinged with an inevitable insecurity about national identity (Baldacchino, 2002, 2009; Mitchell, 2002). Who, exactly, are the Maltese? What are their defining national characteristics, given their proclaimed Phoenician heritage, their unique language, and (yet) their subsequent contact with Arab/North African, Italian and British colonial spheres, and consequent histories of immigration? According to Abela (2006), if there is any common feature of Maltese identity, apart from the obvious attachment to their 'island place', it is the Catholic Church, which provides an overarching umbrella of ontological security for the family of the Maltese people, as well as symbolic visual dominance in the scale of the monumental churches rising over Maltese small towns and villages. Cassar (2000) has shown that, historically, Catholic Christian narratives and imagery have not only played a central role in the construction and maintenance of national identity but also have been deployed to guide debates on morality, welfare and social responsibility – including those relating to migration. Migration links, in both directions, between Malta and Sicily were particularly intense during the sixteenth and seventeenth centuries (Cassar, 2000: 95-120).

Immigrants, therefore, are seen as a potential disruption of this fragile Maltese identity with its unique but beleaguered culture and language. But there are at least two ironies here. The first is to do with the nature of Maltese Catholicism. On the one hand there are very high levels of church attendance and unusually high levels of charity donations to worthy causes, both at home and in the 'Third World'. But when the Third World comes to Malta in the embodiment of dark-skinned immigrants, the charity turns to xenophobia, as noted above. The second irony is the failure to make connections between the Maltese experience of mass emigration and the plight of immigrants arriving in Malta10. One further irony connects the previous two, and this is the role of certain liberal Catholic organizations, such as the Emigrants Commission (originally founded to support Maltese emigrants) and the

Jesuit Refugee Service in acting as providers of important welfare and support services to immigrants and asylum-seekers11 .

The second discursive framing of irregular migration and asylum-seekers concerns the question of numbers; the notion that “we can’t all live on this small island”. It is undeniable that Malta has the highest population density of any EU country: 400,000 on just 316 square kilometres of strategically located but agriculturally infertile limestone rock. Although the annual influx of more than 1 million tourists is welcomed because of the importance of tourism for the economy and because, of course, the tourists go home, the annual arrival of 2,000 ‘illegal’ immigrants who will not, or cannot, return home, is seen as a more vital component in the numbers game. In Malta, it is often pointed out that the impact of 2,000 immigrants on a population of 400,000 is equivalent to the annual arrival of 300,000 boat migrants in Italy or 200,000 in Spain. Or, put another way, the 11,500 refugees and irregular migrants processed since 2002 corresponds to about 1.7 million arriving in the UK, France or Italy (Jones, 2008). A further layer of this discourse reflects Malta’s self-perception as a ‘less-developed country’ in European terms, and therefore ill-equipped to deal with the migrant influx. Here the contrast is drawn with other, more wealthy European countries where it is explicitly acknowledged that migrants have a valuable economic function in the labour market, filling jobs that native workers are unwilling to do, such as in agriculture or construction12 .

This leads to the third, and by far the most complex, challenge: the policy response. This in turn has two dimensions: internal policy measures for the reception, processing, accommodation and integration of migrants and refugees; and the external policy environment of the EU and bilateral relations with sending and transit countries, notably Libya.

What is clear from the few existing studies of the phenomenon of Malta’s ‘new immigration’ (King & Thomson, 2008; Spiteri, 2004; Texiere, 2006; Thomson, 2006) is that the country was unprepared for those who, echoing the biblical shipwreck of St. Paul, washed up on the shores of Malta in the early 2000s13. Although on paper the figures for the eventual confirmation of refugee and humanitarian protection status are rather impressive by southern European standards (Thomson, 2006: 6), the way the irregular immigrants and asylum-seekers have been accommodated has been widely condemned14 . For up to 18 months (now reduced to 12) they are held in overcrowded and poorlyequipped detention camps (which are former military barracks) under the control of soldiers who are trained in warfare not humanitarian work. Subsequently they can be ‘released into the community' which usually means they are transferred to 'open centres' from which they can obtain a work permit and seek work15. There are no real policies or measures for integration and welfare beyond those offered by charitable bodies.

The issue of irregular migration has also had an impact on the domestic political scene, boosting the influence of the right-wing Alleanza Nazzjonali (National Alliance) for whom immigration represents a threat, in their own words, to 'Maltese values of faith, nationhood and family'. Public protests against irregular migration have led to clashes in the main square of the capital Valletta between right-wing campaigners and anti-racist groups. Further rhetorical fuel has been thrown on the fiery debate over immigration by the extremist Imperium Europa group which espouses a white supremacist European credo. Articles and letters in Maltese newspapers further reveal the naïveté of the discussion on immigration, with tired allusions to crime and disease, religious fanaticism and clashes of cultures (Spiteri, 2004). These anti-immigrant movements carry only minimal electoral support, but their voice has enhanced importance in the context of the knife-edge balance of power between Malta's two long-established centrist political parties, the Nationalist Party and the Malta Labour Party (Baldacchino, 2008b).

Malta's external policy challenge involves convincing the EU that it deserves special treatment and assistance due to the disproportionate 'burden' of immigration that it has to bear by virtue of its four undeniable geographical features: small size, dense population, location close to the North African coast, and its huge area of maritime responsibility. At the negotiations for the EU entry, Malta did secure two derogations which reflected a different concern – immigration from other EU countries, such as Italy where there was fear of the immigration of unemployed Sicilians. The two exception clauses were a ban on foreigners purchasing second homes in Malta, and a seven-year moratorium on the free entry of EU workers (with permits only granted in exceptional circumstances for a specific purpose). Less success has been gained in relaxing the rules of the 1990 Dublin Convention, which effectively traps migrants in Malta, against the wishes of the Maltese authorities and of the migrants themselves, who want to move north. And a bilateral agreement with Libya allowing Malta to send irregular migrants back has so far proved elusive, as have requests to the Libyans to control the departure of boats from their shores.

Recent attempts by Italy, Greece, Malta and Cyprus to establish a greater EU commitment to 'burden-sharing' over the issue of Mediterranean 'boat-people' migrants have yielded few concrete results. Winter arrivals in late 2008 on the islands of Lampedusa and Linosa, close to Malta, prompted a strong burst of rhetoric from Italy's interior minister Roberto Maroni; if his pressure on the Libyan leadership, linked to a programme of Italian investment in Libya agreed in August 2008, bears fruit and Gaddafi follows through with the quid pro quo of joint coastal patrols of Libya, then Malta might also 'benefit' from reduced arrivals (Hooper, 2008).

So, does Malta need immigrants? Is there a structural need for migrant workers in the Maltese labour market? One straight answer might be 'Yes', simply because Malta is like any other developed and regulated market economy that is looking to lower production costs. But labour migration is not an absolutely inevitable accompaniment of capitalist success as the case of Finland – one of the most prosperous countries in the EU but with very few labour immigrants – shows (Bartram 2007). For Malta, opinions and evidence differ; we await a thorough study on this question. Certainly there is 'soft' evidence on the presence of a number of immigrant groups beyond the mainly African asylum-seekers and irregular migrants discussed above. There is a Libyan presence which has its roots in the 1970s when prime minister Dom Mintoff's pro-Libyan stance brought Libyan teachers to Malta to teach Arabic in the local schools. As Libya became even more a pariah state in the wake of the 1988 Lockerbie incident, Libyans continued to have visa-free access to Malta until Malta turned towards Europe in the late 1990s. An older-established but small Indian trader community also contributes to Malta's very embryonic multicultural society (Falzon, 2001). Evidence exists for various East European and Middle Eastern migrants working in construction (men) and the hospitality industry (women), and there are Nigerians in the local football teams (Thomson, 2006: 7-8). But in general Malta has a long way to go before it reaches the degree of immigrant penetration into the labour market that has been achieved in other southern EU countries such as Spain, Italy, Greece and Cyprus (King & Thomson, 2008).

Coming back to the central theme of this paper, the Maltese case illustrates a fundamental dilemma at the heart of the relationship between small islands and migration: islands tend to be 'good' at emigration, because that has been their livelihood for generations; but 'bad' at coping with new immigration, certainly if the latter is on a large scale proportionate to the island's population. Other factors also help to explain this anomaly. Partly the situation has to do with the asymmetry in the migratory relationship between islands and the rest of the world: 100,000 Maltese migrants distributed between four big countries (Australia, Canada, the USA, Britain) have little visibility there; but even

10,000 immigrants in Malta would have high visibility. Moreover the changing geopolitical situation of Malta also is important, as in the combination of EU membership with its geographical positionality at the 'underbelly' of the Union, close to migrant-source routes.

# 8. CONCLUSION

This paper has argued for a new generation of island migration studies: one which no longer focuses on emigration, depopulation, island diasporas and return migration, but on islands' new role in global relations and on the changing geopolitics of international migration control. In this conclusion, I revisit some of the key points I have made and try to set them within wider theoretical and methodological debates.

First, I have argued that islands can continue to function as a privileged laboratory for migration studies. As the nature of mobility changes in the 'age of global migration', and as new paradigms reshape the focus of migration researchers in their quest to understand the changing nature and geography of global mobility, the heuristic role of islands as spatial laboratories remains full of (mostly unrealized) potential. This much is consistent with what I have argued in the past (see King, 1999; also Connell & King, 1999). But this view should not pass unchallenged, and therefore needs further nuancing and justification.

Clifford Geertz critiqued the 'spatial laboratory' theme in his 'thick description' introduction to his 1973 classic The Interpretation of Cultures. Later works by anthropologists such as Marcus (1995) and Candea (2007) also debate the issue of the relevance of the local, small-scale place or community. Although those authors rarely refer to the specificity of a small island as such a bounded site, what they say has relevance for the island-as-spatial-microcosm approach16. Marcus' often-cited essay on the emergence of multi-sited ethnography marked the passage from the traditional imaginary of anthropological fieldwork carried out in a single-site location to a more dynamic ethnography contextualized by the structurings of a larger social order, above all the capitalist world system. Hence 'the local' could only be understood – and researched – with reference to observations sited elsewhere in the global system. As tracking strategies he suggested following 'the people' (as in migration and mobility), the 'thing' (as in a commodity chain), the biography (as in life-history studies), or following the metaphor, story or conflict (Marcus, 1995: 106-110).

Second, I have pointed to the new function of islands arising from their strategic location as stepping-stones on routes of irregular migration which are themselves constantly shifting in response to 'official' migration control strategies. We noted how some islands – Malta, Lampedusa and the Canaries were specifically mentioned – acquire new salience as critical spaces in the continually changing geography of irregular migration routes from Africa and the Near East to Europe. Sometimes such islands – particularly the larger, more prosperous ones like Sicily, Majorca or Cyprus – are the end-destinations of migrants, but more often they are places en route to somewhere else: liminal spaces where they may quickly pass through, or perch for a while, like migrating birds, or get stuck (as in Malta), unable to move on, yet unwilling to turn back.

The debate on multi-locality launched by Marcus (1995) also allows us to combine in a fruitful way two key approaches to the island-migration nexus. Candea's (2007) rejoinder to Marcus re-emphasized the importance of the 'bounded location' for ethnographic enquiry, either as a single site or as multiple, discrete locations which have their own consistency and meaning – the village, the neighbourhood, the tribal territory or, one might add, the (small) island. On the other hand such 'traditional' anthropological field-sites are also arbitrary locations which function as heuristic devices which, in turn, are not so much 'ideal types' or closed-system spatial laboratories; but, rather, reflect "actually existing instances, whose messiness, contingency, and lack of an overwhelming coherence or meaning serve as a 'control' for a broader object of study" (Candea, 2007: 180) – in our case, migration. Hierarchy of scale becomes important here, as the small, bounded fieldsite might indeed be a small island which functions, like a rural village, as a 'coherent community'; or the island (Malta, Corsica...) might be large enough to be composed of several small communities yet for there still to be present a marked insular identity amongst its inhabitants.

Marcus' line of argument fits closely with my alternative suggestion for seeing islands not, in this instance, as semi-closed systems or small-scale laboratories for social research, but as critically placed nodes within much broader dynamics linking wider spatial realms and processes, potentially global in scale. By its very nature, migration links islands to other places; indeed migration is the most obvious exemplification of the need for multi-sited research, since there is always a 'here' and a 'there' involved. This, surely, is the attraction of islands for migration researchers (and the attraction of migration for island researchers).

## ACKNOWLEDGEMENTS

This paper is based on a keynote lecture to the conference 'Connecting Worlds: Emigration, Immigration, and Development in Insular Spaces', held at Angra do Heroismo, Azores, 29- 30 May 2008. Thanks to Lucinda Fonseca, Daniela DeBono and Sharon Spiteri for useful discussions around the themes of this paper, and to Island Studies Journal's three anonymous referees for some helpful suggestions which I have tried to respond to. My visits to Malta in 2005 and 2006 were funded by a Social Sciences Small Grant from the Nuffield Foundation (grant SGS/01207/6) which is gratefully acknowledged.

## REFERENCES

1. Abela, A.M. (2006) 'Shaping a National Identity: Malta in the European Union', International Journal of Sociology, Vol. 35, No. 4, pp.10-27.
2. Anon (1964) 'Malta: Migration's Microcosm', International Migration Digest, Vol. 1, No. 2, pp. 211-213.
3. Apostolopoulos, Y. & Gayle, D.J. (eds.) (2002) Island Tourism and Sustainable Development: Caribbean, Pacific and Mediterranean Experiences, Westport CT, Praeger.
4. Baldacchino, G. (2002) 'A Nationless State? Malta, National Identity and the European Union', West European Politics, Vol. 25, No. 4, pp. 191-206.
5. Baldacchino, G. (2004) 'The Coming of Age of Island Studies', Tijdschrift voor Economische en Sociale Geografie, Vol. 95, No. 3, pp. 272-283.

6. Baldacchino, G. (2005) 'Islands: Objects of Representation', Geografiska Annaler, Vol. 87B, No. 4, pp. 247-251.

7. Baldacchino, G. (2007) 'Introducing a World of Islands' in G. Baldacchino (ed.) A World of Islands: An Island Studies Reader, Charlottetown, Canada and Luqa, Malta, University of Prince Edward Island, Institute of Island Studies and Agenda Academic, pp. 1-29.

8. Baldacchino, G. (2008a) 'Studying Islands: On Whose Terms?', Island Studies Journal, Vol. 3, No.1, pp. 37-56.

9. Baldacchino, G. (2008b) 'Immigrants, Tourists and Others from Away: "Come Visit, but Don't Overstay". The Threat of Invasion to Island Societies', Paper to the Conference 'Connecting Worlds: Emigration, Immigration and Development in Insular Spaces', Angra do Heroismo, Azores, 29-30 May 2008.

10. Baldacchino, G. (2009) 'Pangs of Nascent Nationalism from the Nationless State? Eurocoins and Undocumented Migrants in Malta, Post European Union Membership', Nations and Nationalism, Vol. 15, No. 1, pp. 148-165.

11. Baldacchino, G. & Greenwood, R. (eds.) (1998) Competing Strategies of Socio-Economic Development for Small Islands, Charlottetown, Canada, University of Prince Edward Island, Institute of Island Studies.

12. Baldacchino, G. & Milne, D. (eds.) (2000) Lessons from the Political Economy of Small Islands: The Resourcefulness of Jurisdiction, Basingstoke, Macmillan.

13. Bartram, D. (2007) 'Conspicuous by their Absence: Why are there so few Foreign Workers in Finland?', Journal of Ethnic and Migration Studies, Vol. 33, No. 5, pp. 767-782.

14. Basch, L., Glick Schiller, N. & Szanton Blanc, C. (1994) Nations Unbound: Transnational Projects, Postcolonial Predicaments, and Deterritorialized Nation-States, New York, Gordon and Breach.

15. Beck, U. & Beck-Gernsheim, E. (2002) Individualization: Institutionalized Individualism and its Social and Political Consequences, London, Sage.

16. Bertram, G. & Poirine, B. (2007) 'Islands and Political Economy' in G. Baldacchino (ed.) A World of Islands: An Island Studies Reader, Charlottetown, Canada and Luqa, Malta, University of Prince Edward Island, Institute of Island Studies and Agenda Academic, pp. 325-377.

17. Bertram, G. & Watters, R.F. (1985) 'The MIRAB Economy in the South Pacific Microstates', Pacific Viewpoint, Vol. 26, No. 3, pp. 497-520.

18. Bertram, G. (2006) 'The MIRAB Model in the Twenty-First Century', Asia Pacific Viewpoint, Vol. 47, No. 1, pp. 1-14.

19. Biagini, E. & Hoyle, B. (eds.) (1999) Insularity and Development: International Perspectives on Islands, London, Pinter.

20. Bianchi, R. (2000) 'Migrant Tourist-Workers: Exploring the "Contact Zones" of PostIndustrial Tourism', Current Issues in Tourism, Vol. 3, No. 2, pp. 107-137.

21. Bourdieu, P. (1986) 'The Forms of Capital' in J.G. Richardson (ed.) Handbook of Theory and Research for the Sociology of Education, New York, Greenwood Press, pp. 241-258.

22. Boyle, P., Halfacree, K. & Robinson, V. (1998) Exploring Contemporary Migration, London, Longman.

23. Braudel, F. (1972) The Mediterranean and the Mediterranean World at the Time of Philip II, London, Methuen.

24. Briguglio, L. (1995) 'Small Island Developing States and their Economic Vulnerabilities', World Development, Vol. 23, No. 9, pp. 1615-1632.

25. Briguglio, L., Archer, B., Jafari, J. & Wall, G. (eds.) (1996a) Sustainable Tourism in Islands and Small States, Vol. 1: Issues and Policies, London, Pinter.

26. Briguglio, L., Butler, R.W., Harrison, D. & Leal Filho, W. (eds.) (1996b) Sustainable Tourism in Islands and Small States, Vol. 2: Case Studies, London, Pinter.

27. Brookfield, H. (1990) 'An Approach to Islands' in W. Beller, P. d'Ayala & P. Hein (eds.) Sustainable Development and Environmental Management of Small Islands, Paris, Parthenon, pp. 23-33.

28. Brunhes, J. (1920) Human Geography, London, Harrap. Candea, M. (2007) 'Arbitrary Locations: In Defence of the Bounded Field-site', Journal of the Royal Anthropological Institute, Vo. 13, No.1, pp.167-184.

29. Carling, J. (2002) 'Migration in the Age of Involuntary Immobility: Theoretical Reflections and Cape Verdean Experiences', Journal of Ethnic and Migration Studies, Vol. 28, No. 1, pp. 5-42.

30. Carling, J. (2007a) 'Migration Control and Migrant Fatalities at the Spanish-African Borders', International Migration Review, Vol. 41, No. 2, pp. 316-343.

31. Carling, J. (2007b) 'Unauthorized Migration from Africa to Spain', International Migration, Vol. 45, No. 4, pp. 3-37.

32. Casado-Díaz, M.A., Kaiser, C. & Warnes, A.M. (2004) 'Northern European Retired Residents in Nine Southern European Areas: Characteristics, Motivations, and Adjustment', Ageing and Society, Vol. 24, No. 3, pp. 353-381.

33. Cassar, C. (2000) Society, Culture and Identity in Early Modern Malta, Malta, Mireva Publications.

34. Castles, S. & Miller M.J. (2003) The Age of Migration, Basingstoke, Palgrave Macmillan.

35. Clark, E., Johnson, K. Lundholm, E. & Malmberg, G. (2007) 'Island Gentrification and Space Wars' in G. Baldacchino (ed.) A World of Islands: An Island Studies Reader, Charlottetown, Canada and Luqa, Malta, University of Prince Edward Island, Institute of Island Studies and Agenda Academic, pp. 481-512.

36. Collyer, M. (2007) 'In-between Places: Trans-Saharan Transit Migrants in Morocco and the Fragmented Journey to Europe', Antipode, Vol. 39, No. 4, pp. 668-690.

37. Connell, J. & King, R. (1999) 'Island Migration in a Changing World' in R. King & J. Connell (eds.) Small Worlds, Global Lives: Islands and Migration, London, Pinter, pp. 1- 26.

38. Connell, J. (1988) Sovereignty and Survival: Island Microstates in the Third World, Sydney, University of Sydney, Department of Geography, Research Monograph, No. 3.

39. Connell, J. (1994) 'The Cayman Islands: Economic Growth and Immigration in a British Colony', Caribbean Geography, Vol. 5, No. 1, pp. 51-66.

40. Connell, J. (2007) 'Migration' in G. Baldacchino (ed.) A World of Islands: An Island Studies Reader, Charlottetown, Canada and Luqa, Malta, University of Prince Edward Island, Institute of Island Studies and Agenda Academic, pp. 455-481.

41. Defoe, D. (1719/1980) Robinson Crusoe, London, Penguin.

42. Delia, E.P. (1982) 'The Determinants of Modern Maltese Emigration', International Migration, Vol. 20, No.1-2, pp. 11-25.

43. Dommen, E. & Hein, P. (eds.) (1985) States, Microstates and Islands, London, Croom Helm.

44. Falzon, M.A. (2001) 'Origins and Establishment of the Indian Business Community in Valletta', Bank of Valletta Review, No. 24, pp. 73-92.

45. Falzon, M.A. (2007) 'God Protect me from my Friends: Prelates, Politicians, and Social Welfare in Contemporary Malta', Journal of Mediterranean Studies, Vol. 17, No.1, pp. 47- 72.

46. Felgentreff, C. (1999) 'Spatial Population Mobility as Social Interaction: A Fijian Island's multi-local Village Community' in R. King & J. Connell (eds.) Small Worlds, Global Lives: Islands and Migration, London, Pinter, pp.255-276.

47. Geertz, C. (1963) Agricultural Involution: The Process of Ecological Change in Indonesia, Berkeley CA, University of California Press.

48. Geertz, C. (1973) The Interpretation of Cultures, New York, Basic Books.

49. Glick Schiller, N. & Fouron, G. (1998) 'Transnational Lives and National Identities: The Identity Politics of Haitian Immigrants' in M.P. Smith & L.E. Guarnizo (eds.) Transnationalism from Below, New Brunswick NJ, Transaction Publishers, pp. 130-161.

50. Gössling, S. & Wall, G. (2007) 'Island Tourism' in G. Baldacchino (ed.) A World of Islands: An Island Studies Reader, Charlottetown, Canada and Luqa, Malta, University of Prince Edward Island, Institute of Island Studies and Agenda Academic, pp.429-453.

51. Gössling, S. (ed.) (2003) Tourism and Development in Tropical Islands: Political Ecology Perspectives, Cheltenham, Edward Elgar.

52. Guarnizo, L.E. (1994) 'Los 'Dominican Yorks': The Making of a Binational Society', Annals of the American Academy of Political and Social Science, Vol. 533, pp. 70-86.

53. Guarnizo, L.E. (1997) 'The Emergence of a Transnational Social Formation and the Mirage of Return amongst Dominican Transmigrants', Identities, Vol. 4, No. 2, pp. 281- 322.

54. Hamood, S. (2008) 'EU-Libya Cooperation on Migration: A Raw Deal for Refugees and Migrants?', Journal of Refugee Studies, Vol. 21, No. 1, pp. 19-42.

55. Hall, S. (1996) 'Who needs Identity?' in S. Hall & P. Du Gay (eds.) Questions of Cultural Identity, London, Sage, pp.1-17.

56. Hay, P. (2006) 'A Phenomenology of Islands', Island Studies Journal, Vol. 1, No. 1, pp. 19-42.

57. Hintjens, H.M. & Newitt, M.D.D. (eds.) (1992) The Political Economy of Small Tropical Islands: The Importance of Being Small, Exeter, University of Exeter Press.

58. Hooper, J. (2008) 'Thousands more Migrants reach Italy's Shores', The Guardian, 30 December 2008, p.17.

59. Itzigsohn, J., Cabral, C.D., Medina, E.H. & Vazquez, O. (1999) 'Mapping Dominican Transnationalism: Narrow and Broad Transnational Practices' Ethnic and Racial Studies, Vol. 22, No. 2, pp. 316-339.

60. Jones, A. (2008) 'Malta: Where Hysteria is no Answer to the Plight of Refugees', The Guardian, 30 December 2008, p.17.

61. Jones, H.R. (1973) 'Modern Emigration from Malta', Transactions of the Institute of British Geographers, Vol. 60, pp.101-120.

62. Kaye, R. (2001) '"Blaming the Victim": "An Analysis of Press Representation of Refugees and Asylum-seekers in the United Kingdom in the 1990s' in R. King & N. Wood (eds.)

63. Media and Migration: Constructions of Mobility and Difference, London, Routledge, pp. 53-70.

64. King, R. (1979) 'Post-war Migration and Return Migration in Malta', European Demographic Information Bulletin, Vol. 10, No. 3, pp. 108-128.

65. King, R. (1993) 'The Geographical Fascination of Islands' in D.G. Lockhart, D. DrakakisSmith & J.A. Schembri (eds.) The Development Process in Small Island States, London, Routledge, pp.13-37.

66. King, R. (1996) 'Migration in a World Historical Perspective' in J. van den Broeck (ed.) The Economics of Labour Migration. Cheltenham, Edward Elgar, pp. 7-75.

67. King, R. (1999) 'Islands and Migration' in E. Biagini & Hoyle, B. (eds.) Insularity and Development: International Perspectives on Islands, London, Pinter, pp. 93-115.

68. King, R. (2002) 'Towards a New Map of European Migration' International Journal of Population Geography, Vol. 8, No. 2, pp. 89-106.

69. King, R. (2007) A History of Migration, London, Marshall Editions. King, R. (2008) Review of G. Baldacchino (ed.) A World of Islands: An Island Studies Reader, Island Studies Journal, Vol. 3, No. 2, pp. 131-134.

70. King, R. & Connell, J. (eds.) (1999) Small Worlds, Global Lives: Islands and Migration, London, Pinter.

71. King, R. & Strachan, A. (1980) 'The Effects of Return Migration on a Gozitan Village', Human Organization, Vol. 39, No. 2, pp. 175-179.

72. King, R. & Thomson, M. (2008) 'The Southern European Model of Immigration: Do the cases of Malta, Cyprus and Slovenia fit?', Journal of Southern Europe and the Balkans, Vol. 10, No. 3, pp. 265-291.

73. Lawrence, D.H. (1926/1982) The Man Who Loved Islands, London, Penguin.

74. Lockhart, D.G. & Drakakis-Smith, D. (eds.) (1997) Island Tourism: Trends and Prospects, London, Pinter.

75. Lockhart, D.G., Drakakis-Smith, D. & Schembri, J.A. (eds.) (1993) The Development Process in Small Island States, London, Routledge.

76. Lowenthal, D. (1992) Small tropical islands: a general overview, in H.M. Hintjes & M.D.D. Newitt (eds.) The Political Economy of Small Tropical Islands, Exeter, Exeter University Press, pp. 18-29.

77. Marcus, G. (1995) 'Ethnography in/of the World-system: The Emergence of Multi-Sited Ethnography', Annual Review of Anthropology, Vol. 24, No. 1, pp. 95-117.

78. Mazetti, E. (2007) 'Radiografia di un Organismo Geografico Complesso: L'Isola di Capri', Rivista Geografica Italiana, Vol. 114, No. 4, pp. 593-619.

79. McCall, G. (1994) 'Nissology: The Study of Islands', Journal of the Pacific Society, Vol. 17, No. 2-3, pp. 1-14.

80. McCall, G. (1996) 'Clearing Confusion in a Disembedded World: The Case for Nissology', Geographische Zeitschrift, Vol. 84, No. 2, pp. 74-85.

81. Mitchell, J. (2002) Ambivalent Europeans: Ritual, Memory and the Public Sphere in Malta, London, Routledge.

82. Price, C.A. (1954) Malta and the Maltese: A Study in Nineteenth-Century Migration, Melbourne, Georgian House.

83. Ratter, B.M.W. & Sandner, G. (1996) 'Small Islands, Large Questions', Geographische Zeitschrift, Vol. 84, No. 2, pp. 63-66.

84. Ravenstein, E.G. (1885) 'The Laws of Migration – I', Journal of the Royal Statistical Society, Vol. 48, No. 2, pp. 167-235.

85. Ravenstein, E.G. (1889) 'The Laws of Migration – II', Journal of the Royal Statistical Society, Vol. 52, No. 2, pp. 241-305.

86. Read, R. (2004) 'The Implications of Increasing Globalization and Regionalism for the Economic Growth of Small Island States', World Development, Vol. 32, No. 2, pp. 365- 378.

87. Robertson, G. (2007) 'Resources for Island Studies' in G. Baldacchino (ed.) A World of Islands: An Island Studies Reader, Charlottetown, Canada and Luqa, Malta, University of Prince Edward Island, Institute of Islands Studies and Agenda Academic, pp. 539-577.

88. Salvá-Tomás, P.A. (2002) 'Tourist Development and Foreign Immigration in the Balearic Islands', Revue Européenne des Migrations Internationales, Vol. 18, No. 1, pp. 87-101.

89. Schneider, P. & Schneider, J. (1976) Culture and Political Economy in Western Sicily, New York, Academic Press.

90. Semple, E.C. (1911) Influences of the Geographic Environment, London, Constable.

91. Sheller, J. & Urry, J. (2006) 'The New Mobilities Paradigm', Environment and Planning A, Vol. 38, No. 2, pp. 207-226.

92. Skeldon, R. (2006) 'Interlinkages between Internal and International Migration and Development in the Asian Region', Population, Space and Place, Vol. 12, No. 1, pp. 15- 30.

93. Smith, A. (2006) Colonial Memory and Postcolonial Europe: Maltese Settlers in Algeria and France, Minneapolis MN, University of Indiana Press.

94. Spiteri, S. (2004) Treasure Island: A journey of Self-Discovery with the Help of Asylumseekers and Refugees in Malta, Cardiff, University of Cardiff, School of Journalism, MA dissertation.

95. Texiere, F. (2006) At the Gate of Fortress Europe: Irregular Immigration and Malta. Rennes, France, Institute for Political Studies Sciences Po Rennes (thesis).

96. Thomson, M. (2006) Migrants on the Edge of Europe: Perspectives from Malta, Cyprus and Slovenia, Brighton, University of Sussex, Sussex Migration Working Paper, 35.

97. Urry, J. (2000) Sociology beyond Society: Mobilities for the 21st Century, London, Sage.

98. Vertovec, S. & Cohen, R. (eds.) (1999) Migration, Diasporas and Transnationalism, Cheltenham, Edward Elgar.

99. Vidal de la Blache, P. (1926) Principles of Human Geography, London, Constable.

100. Vine, D. (2004) 'War and Forced Migration: The United States Military Base at Diego García', International Migration, Vol. 42, No. 3, pp. 111-143.

101. Warnes, A.M. & Patterson, G. (1998) 'British Retirees in Malta: Components of the Crossnational Relationship', International Journal of Population Geography, Vol. 4, No. 3, pp.108-128.

102. Warrington, E. & Milne, D. (2007) 'Island Governance' in G. Baldacchino (ed.) A World of Islands: An Island Studies Reader, Charlottetown, Canada and Luqa, Malta, University of Prince Edward Island, Institute of Island Studies and Agenda Academic, pp. 379-427.

103. Williams, A.M. & Fonseca, M.L. (1999) 'The Azores: Between Europe and North America' in R. King & J. Connell (eds.) Small Worlds, Global Lives: Islands and Migration, London, Pinter, pp. 55-76.

10

# ECOTOURISM IN DOMINICA: STUDYING THE POTENTIAL FOR ECONOMIC DEVELOPMENT, ENVIRONMENTAL PROTECTION AND CULTURAL CONSERVATION

**Vanessa Slinger-Friedman**

Department of Geography and Anthropology Kennesaw State University Georgia, USA.

## ABSTRACT

Over the last 20 years, the government of the small Caribbean island nation of Dominica has pursued the development of ecotourism on the island. The hope is that this industry will promote dispersed economic development while providing environmental protection and cultural conservation. However, not enough has been done to determine whether or not the industry is achieving the desired results. To this end, this study explores the growth of the tourism industry on Dominica and shows that, while not perfect, ecotourism is providing some definite benefits to the island in this regard.

## KEYWORDS

Caribs cultural conservation, Dominica, economic development, ecotourism, environmental protection, islands

# 1. INTRODUCTION

Since the 1990s, Dominica has experienced a significant decline in the export of bananas, its principal crop, due to weather conditions and loss of preferential trade access in European markets. In response, the Dominican government has pursued the development of the ecotourism industry on the island. Ecotourism is considered to have strong potential for economic growth and diversification because of the island's abundance of mountains, streams, forests, and spectacular flora and fauna, as well as its unique indigenous culture.

The hope of many in academia, national governments, and the tourism industry is that ecotourism will be the antithesis to conventional mass tourism. Historically, sun, sea, and sand tourism in the Caribbean has been linked to large scale, foreign-owned and run facilities, the destruction of the environment, rising imports, the repatriation of profits, polarized development, and the loss of cultural identity (Archer, 1998; Brohman, 1996; Blommestein, 1995; Britton, 1982; Caribbean Tourism Organization [CTO], 1999; Klak & Flynn, 2008). On the other hand, ecotourism, which involves visitors enjoying the natural environment and culture of a location, has expanded in meaning to incorporate tourist education, environmental conservation, benefits to the local population, and the sustainability of local cultures. However, a review of studies on ecotourism in other wellknown ecotourism locations - such as the Galápagos Islands, Costa Rica, Belize and Kenya - have shown that, while some small, locally-inspired projects are promising, large, foreign-owned establishments have generally dominated the tourism industry and are favoured by local governmental policies (Honey, 1999, 2008; Place, 1998; Weaver, 1998; Woods et al., 1998).

The study of ecotourism in Dominica is important for several reasons. First of all, the island, with strong government backing in the form of policies and planning, is trying to develop a deliberate and comprehensive ecotourism industry. Second, Dominica has a land area of 290 square miles (750 km2 ) and is a small enough island for the entire country to be studied and for any policy impacts to be more readily observable. Third, a literature review shows that research on ecotourism in small island developing nations remains lacking. Finally, and on the more practical side, ecotourism is growing in importance to many Caribbean islands as they lose preferential access to markets for their agricultural produce and recognize the negative effects of traditional tourism.

With bananas as the principal crop, agriculture is the economic mainstay of Dominica (Bonnerjea & Weir, 1996). Bananas were introduced from the Canary Islands in 1516, and since the mid-1950s have been so important to the country's economy that they were nicknamed 'Green Gold' (Latin American Bureau, 1987). However, the agricultural sector of Dominica has been vulnerable to hurricanes, fluctuations in world market prices, and especially to the decline in preferential trade access to European markets. Due to these factors, Dominica's banana exports decreased from US$38 million in 1988 to less than US$5 million in 2003 (Euromonitor International, 2007). Concurrently, the number of banana farmers fell from 4,366 in 1995 to 1,400 in 2003 (Encyclopedia of the Nations, 2008; Progressive Policy Institute, 2004). As a result, the Dominican government has increasingly focused on economic diversification, and much of this effort has been centered on the development of ecotourism.

Worldwide, tourism has long been used as a strategy for promoting economic development in developing countries by providing foreign capital earnings, employment, and linkages to other sectors of their economies (Asiedu, 1997; Brohman, 1996). Since the 1970s, tourism based on the availability of the sun, sea and white sand beaches has been a key element in the economic development plans for many Caribbean islands. Tourism in the insular Caribbean has increased its market share and out-performed the industry worldwide (Duval, 2004; McElroy, 2004). However, the projected economic benefits of tourism in the Caribbean have been mixed, and even proven illusory. Apart from, or instead of, the desired positive impacts, traditional tourism has often been associated with foreign ownership, profit repatriation, accentuation of regional disparities, and environmental and cultural destruction (Klak, 1998).

In this most tourism intensive region of the world, Dominica captures less than 1% of tourists to the Caribbean (WTO, 2009). However, Dominica represents an alternative form of tourism experience demanded by more and more visitors from key international markets. The demand for ecotourism has grown simultaneously with the recognition of the environmental concerns associated with the conventional mass tourism most common in the insular Caribbean.

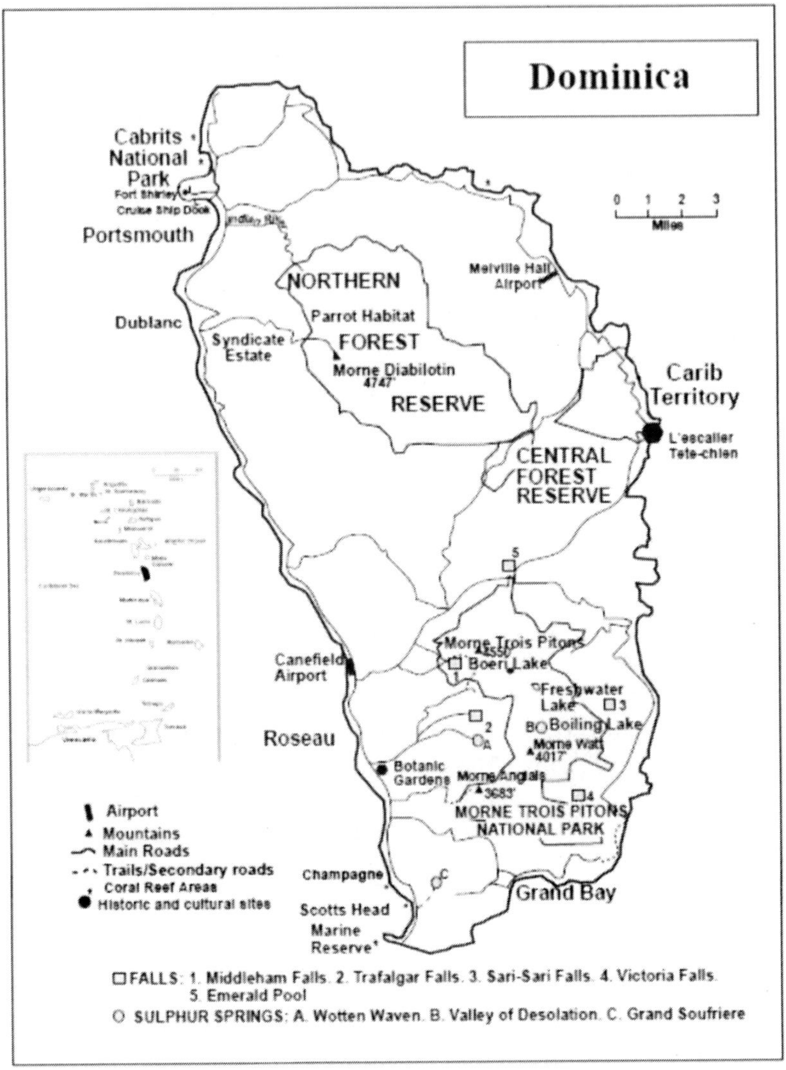


**Figure 1.** Map of Dominica

Whereas many other Caribbean islands turned to the traditional 'sun, sea, and sand' type of tourism, Dominica's mountainous terrain, rocky shores, and black sand beaches deterred its development in this direction (Figures 1 and 2). Tourists to Dominica were scarce until the 1980s, when the government decided to capitalize on the island's unique physical and cultural assets by promoting Dominica as an ecotourism destination (Hulme, 2007). This type of tourism seemed ideal due to Dominica's abundant tropical rainforests, hot sulphur springs, waterfalls, whale watching opportunities, the Boiling Lake (the second largest boiling fumarole in the world), and the presence of the Caribs, the last remaining group of indigenous people in the Caribbean. Other biological assets identified as tourist attractions are the thousands of species of flowering plants, including 74 species of orchid and 200 ferns and over 5,000 species of vascular plants. There is also a large bat and bird population, including the Sisserou (Imperial) and Jaco (Red-necked) parrots, both of which are endangered and endemic to Dominica (Halcrow, 2003; Honychurch, 1991). The seemingly endless supply of rainwater also feeds the numerous lakes and rivers on the island, some of which are navigable and full of interesting biophysical features. The best example of this can be found on the Indian River, where boat trips take tourists through winding, mangrove- fringed waters.

**Figure 2.** Steep coastlines and volcanic black sand beaches of Castle Bruce on Dominica's North-east coast. (Photo taken by the author)

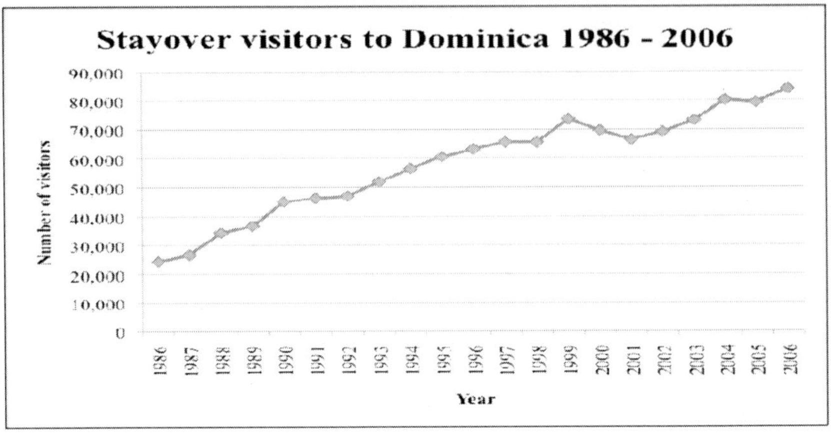


**Figure 3.** Stayover Visitors in Dominica: 1986 -2006.

Dominica might have initially started out as a "circumstantial" alternative tourism destination (Weaver, 1991). Its lack of white sand beaches and insufficient flat land for an international airport limited its involvement in traditional tourism. However, Weaver (1991) points out a shift in perception by the Dominican government with the realization that the island's natural resource base could be an asset to tourism development rather than a liability. Over the last three decades, the Dominican Government's pursuit of a deliberate alternative tourism strategy is reflected in numerous tourism plans and policies that have been commissioned by or on behalf of the Dominican government in order to formulate a national tourism policy (Ministry of Tourism and National Development Corporation (NDC), 2005; Weaver, 1991), and Dominica's share of tourist arrivals to the Caribbean has been growing steadily. For the period of 1989 to 1993, Dominica had the highest annual increase in visitor expenditure among the Caribbean islands (Hypolite et al., 2002). The Dominican tourism sector contributed an estimated US$43 million in direct visitor expenditures in 2003 (Ministry of Tourism and NDC, 2005). Since 1986 stayover tourist arrivals to the island have grown annually by almost 7% (Figure 3), and they generate as much as 88% of tourism revenues for the island when compared to cruise ship visitors (Ministry of Tourism and NDC, 2005).

Despite the Dominican government's ecotourism policy focus, the number of cruise ship visitors to Dominica has also grown significantly, and now far exceeds the number of stayover visitors (Figure 4). Most researchers consider cruise ship tourism to run contrary to the goals of ecotourism because it is often associated with limited local economic impact and potential environmental degradation caused by a large quantity of tourists visiting environmentally sensitive areas (Honey, 2008). The Government of Dominica has dealt with the apparent contradiction between promoting stayover tourism and courting cruise tourism by developing and promoting multiple terrestrial visitation sites for cruise ship tourists. Their goal is to keep the pressure low at any one particular site while obtaining dispersed regional development on the island. To this end, twenty heritage and natural attractions were added in the process (Johnson, 2006). Additionally, the government hopes that cruise ship passengers who experience Dominica for only a few hours will potentially return to the island as stayover tourists1 .

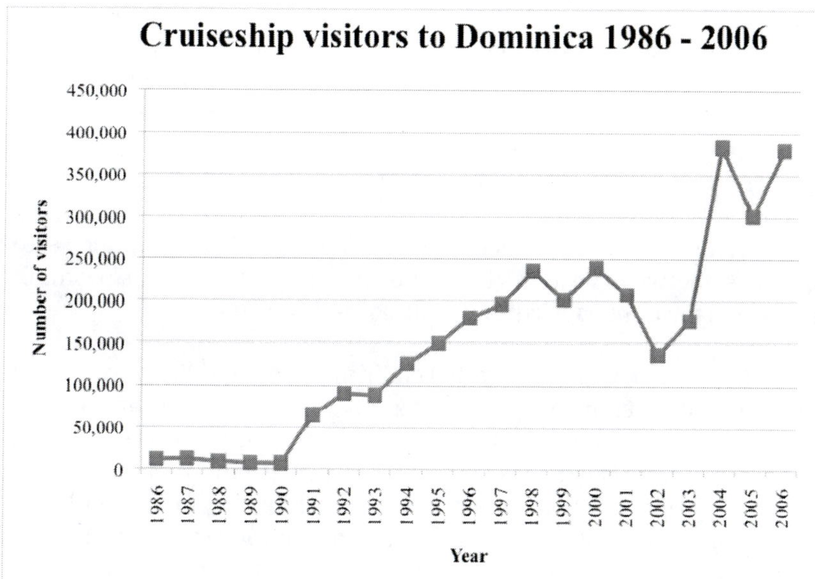


**Figure 4.**Cruise ship visitor arrivals to Dominica 1986 to 2006.

## 2. THE STUDY OF ECOTOURISM ON DOMINICA

Fieldwork and secondary data collection was carried out between 1999 and 2008 to see if ecotourism in Dominica is having the desired results of local involvement, dispersed regional development, linkages with other sectors of the economy, and environmental and cultural preservation.

In 2000 and 2004, fieldwork was carried out on Dominica by the author to complete survey questionnaires with individuals from accommodation facilities, restaurants, tour companies, tour guides, and craft sellers from around Dominica2 . In all, 75 accommodation facilities, 21 restaurants, 43 craft sellers, 5 dive shops, 18 tour companies and 20 guides, 26 taxi operators, and 56 farmers, fishermen, and hucksters (local term for middlemen) completed surveys and answered questions relating to tourism in Dominica. The hotels and dive shops surveyed represented roughly 68% and 70%, respectively, of these types of facilities on the island. The total size of the other groups surveyed is unknown because there typically is no formal organization to which they belong. Even where there are formal groups, many people chose not to belong to these organizations, making it difficult to determine exact participation in the tourism industry. However, based on the small size of the island and the time spent doing detailed studies of specific individuals and communities, I feel that the data collected portrays an accurate picture of tourism on Dominica. Interviews with farmers and hucksters who are presently involved in servicing hotels and restaurants with food, as well as hotel and restaurant owners themselves, revealed participation in the industry and inter-sectoral linkages between ecotourism and other parts of the Dominican economy, including agriculture, fishing, and forestry.

Secondary data was also gathered from 2000 to 2008 from official government documents, other non-government organizations, and a literature review of current and relevant academic studies. This study looked at three main research areas: (1) creation of employment, economic opportunities, and infrastructural development; (2) linkages formed between tourism and other sectors of the Dominican economy; and (3) conservation of the environment and local indigenous culture.

# 3. RESEARCH RESULTS

## 3.1 Economic Impacts

The first research area was the impact of ecotourism on employment creation, economic opportunities, and infrastructural development. Direct employment in the Dominican tourism industry is estimated to have grown from two hundred people in 1974 to roughly 2,5003 or approximately 10.3% to the total work force in 2005 (U.S. Department of State, 2009; CHL, 2006). Tourism's contribution to Dominica's gross domestic product ranges from 10% to 25%; the higher estimates take into account a multiplier effect that factors in indirect employment generated in such industries as agriculture and fishing. Tourism also generated over 30% of the country's foreign exchange: three times the earnings from bananas in 2001 (Halcrow Group Ltd., 2003).

A 1993 Caribbean Tourism Organization survey indicated that the majority of tourists coming to Dominica do so to enjoy its natural beauty and tranquility (CTO, 1993). This was evident by the importance that the respondents attached to sightseeing, peace and quiet, and the interest shown in natural history, including the Carib Territory. In addition, a 1995/96 survey of stayover tourists carried out by a consultant for the Dominican government showed that almost ninety-three percent of visitors had already visited a nature site (Westbrook et al., 1997).

A stayover visitor survey carried out in 1995/1996 (Westbrook et al.,1997), found that expenditures per person were highest for visitors who were coming to Dominica primarily with an interest in nature, with an average of US$883.50 when compared to an average of US$576.90 for those on a general vacation. Another factor adding to the potential increased local value of Dominica tourism is the fact that pre-paid packages are seldom used by visitors coming to Dominica. Four out of five visitors make their own travel arrangements (CTO, 1993).

This study found that a diverse number of employment opportunities are being created in the ecotourism industry in Dominica, including positions in the accommodation sector, construction, food preparation, local craft production and sales, agriculture, fishing, park maintenance, diving, and touring in both the rural and urban areas of Dominica (Figure 5). These opportunities ranged from manual work to managerial and ownership positions. Based on the research, there is a predominance of Dominicans at every level of employment in the tourism industry. Dominicans filled 96% of all positions in the tourism industry, including 93% of professional and managerial positions in the same industry (Table 1). In addition, 92% of all survey respondents (N = 264) said that they did not have to resettle in order to be involved in the ecotourism industry: this suggests that ecotourism is creating employment opportunities that are regionally diverse. Overall, information gleaned from interviews indicated that respondents did not have to travel a significant distance to their tourism job. Economic opportunities through tourism are also being created in the accommodation sector, diving facilities, and craft shops.

**Table 1.** Local versus Foreign Employment in the Dominican Tourism Industry (N = 119)

| Data | Percent |
|---|---|
| Positions in the tourism industry filled by Dominicans | 96 |
| Professional/managerial positions in tourism industry filled by Dominicans | 93 |
| Would prefer to hire locals as employees in their business | 100 |

Surveys show that increases in income since being involved in the tourism industry vary greatly and not all respondents were willing to provide detailed financial information. However, over three quarters of the respondents said that they experienced an increase in their monthly or yearly income since being involved in the tourism industry. While hotel owners were not willing to give specific details, some tour guides reported annual incomes of about US$3,350, some dive instructors reported annual earnings of roughly US$4,500, and someone in the craft shop business reported annual earnings of close to US$6,700. All of these earnings exceeded the per capita GDP of Dominica.

Moreover, almost half of the hotel respondents mentioned adopting behavior in response to changes in tourism demand, including slightly reducing staff numbers or closing the business to take a vacation. A large number of total respondents (86%) also stated that they had alternative sources of income other than tourism revenue. These alternative sources of income, often supported by occupational multiplicity; they include small gardens, remittances from abroad, access to pensions, and public sector employment.

While the strip of overnight accommodation in and around the urban area of Dominica's West points to a slightly higher concentration of units in this area, the overall pattern of units in Dominica is dispersed throughout the country (Table 2). Additionally, the concentration of rooms in the West urban area has declined from 49% in 2000 to 44% in 2008. Analysis of craft shops and vendors on the island also showed a widespread distribution, and although there is a slightly higher distribution of dive shops on the leeward (west) side of the island, this is due to reduced wind, drier weather, and a more gently sloping coastline producing better diving conditions.

**Figure 5.** Local Tour Guide at the Boiling Lake in the Valley of Desolation. (Photo: taken by author)

In 2007, hotels accounted for just 29% of the total number of travel accommodations on the island (Euromonitor International, 2008). This corresponds with the Government's focus on ecotourism, which can come with a preference by visitors for guesthouses and lodges. Weaver (1991) suggests that the potentially allocentric character of tourists to Dominica could make small-scale accommodations more desirable as these tourists seek a more active and authentic local experience. Local ownership predominates in Dominican tourism. Dominicans own the majority of tourism businesses such as accommodation facilities, dive shops and craft shops. Local ownership of accommodation facilities increased from 62% in 1991 to 88% in 2004 (Table 3). Also, no pattern exists of diminished local ownership with increased size of hotel, as can be found on many other islands with conventional tourism (Table 3). Ownership by foreigners without Dominican citizenship tends to be by individuals who have lived in Dominica over ten years or have married Dominicans, and many of these individuals consider Dominica to be their home. Most of these foreigners are not absentee owners, but actually live in and run the facilities themselves. Therefore, it is safe to assume that repatriation of profits from tourism in these cases will be limited. The diverse regional development of ecotourism projects and facilities on the island has led to widespread opportunities for employment throughout Dominica.

**Table 2.**Regional Distribution of Dominican Tourism Accommodation Units, 2008

| Region | No. of facilities | Percentage of total facilities | No. of rooms | Percentage of rooms |
|---|---|---|---|---|
| North | 38 | 27 | 151 | 18 |
| South | 21 | 15 | 111 | 13 |
| West, urban, semi-urban | 42 | 30 | 381 | 44 |
| East | 18 | 13 | 119 | 14 |
| Interior | 21 | 15 | 96 | 11 |
| TOTAL | 140 | 100 | 858 | 100 |

**Table 3.**Ownership and Scale of Accommodation Facilities in Dominica (N = 75).

| Size of facility # units | 100% Dominican | % total | Foreign/ Dominican | Foreign with Dominican citizenship | 100% foreign | Total |
|---|---|---|---|---|---|---|
| <5 | 21 | 96 | 0 | 1 | 0 | 22 |
| 5 to 9 | 21 | 91 | 0 | 2 | 0 | 23 |
| 10 to 24 | 18 | 78 | 1 | 3 | 1 | 23 |
| Over 25 | 6 | 86 | 0 | 0 | 1 | 7 |
| TOTAL | 66 | 88 | 1 | 6 | 2 | 75 |

## 3.2 Sectoral Linkages

The second research statement looked at the linkages formed between tourism and other sectors of the Dominican economy. Results show that tourism in Dominica appears to be complementing and linking to other previously established Dominican industries such as agriculture, fishing, and forestry by creating an added demand for goods and services from these industries. For instance, one facility in the southeast of the island, Jungle Bay resort, obtained 90% of its building material from local sources. Furthermore, 90% of food prepared in the kitchen at this resort is obtained from the local community through a group of 30 farmers (including former banana workers) and other vendors (Sam Raphael (owner), personal communication, 2008). Survey output indicated that two-thirds of the produce and materials used in tourism facilities in Dominica are made locally, and more than half of these facilities' budgets are spent on obtaining local products. One reason that the quantity of money spent on local products is only slightly higher than money spent on imported products is because items from abroad, including butter, beef, flour, and salt, tend to be more expensive (Table 4).

**Table 4.**Use of Locally made Products in the Dominican Tourism Industry (N = 144)

| Data | Percent |
|---|---|
| Amount of locally made products used in providing their tourism service | 66.3 |
| Amount of foreign products used in providing their tourism service | 33.7 |
| Amount of facility budget that goes towards obtaining locally made products for providing their tourism service | 51.4 |
| Amount of the facility budget that goes towards obtaining foreign products for providing their tourism service | 48.6 |

## 3.3 Environmental and Cultural Impacts

The final research statement considers ecotourism's impact on both the environment of Dominica and on local indigenous culture. The government has created land and marine parks to protect the rainforest and reefs of Dominica. Through the establishment of national parks, such as the Morne Trois Pitons National Park in 1975, the Morne Diablotin National Park in 2000, and the Soufriere/Scott's Head Marine Research in 2001, over 20% of the island has been allocated to national parks and protected areas (Patterson & Rodriguez, 2003). Additionally, such legislation as the Forestry and Wildlife Act, the Fisheries Act, and Beach Control Act, has been instituted to protect the island's natural environment. People involved in the Dominican tourism industry overwhelming see nature as both the main attraction for visitors and the resource base for their own livelihoods (Table 5). Tourism businesses also consider the maintenance of the natural environment as a high priority. This view has led to claims of conservation initiatives by 85% of the individuals in the ecotourism industry surveyed (Table 5).

**Table 5.** Attitudes of People in Tourism towards Conservation of Local Environment (N = 264)

| Attitude | Percent |
|---|---|
| Feel that nature is the major reason why visitors come to Dominica | 100 |
| Rely on the surrounding environment for their living | 93.6 |
| Contribute themselves to conserving or improving their surrounding environment | 85 |
| Have done more towards conserving or improving their surrounding environment since being involved in tourism | 77 |
| Feel that more could be done by the Government to conserve Dominica's natural environment for tourism | 85.6 |
| Feel that more could be done by individuals toward conserving Dominica's natural environment for tourism | 47 |

While only a minority of the respondents mentioned local awareness and conservation actions as a benefit of ecotourism, people involved in the Dominican tourism industry see nature as both the attraction for visitors and the resource base for their livelihoods. Tourism business people appear to fully understand their reliance on the environment.

Craft makers depend on the environment for materials; farmers for a good agriculture base; and tour guides, dive shops, restaurants, and hotels for sites of interest for visitors. Tourism businesses see maintenance of the environment as a high priority. This view has led to many personal conservation actions by hotel and restaurant owners, tour guides, dive shops, and craft people. Many individuals in the tourism business have incorporated or increased their conservation practices since being involved in Dominican tourism. Measures taken by these individuals include: community and individual beach and river clean-ups, placing boulders to protect against erosion, growing and using organic produce, using Green Globe recommended cleaning products (such as vinegar to clean furniture), educating locals and tourists on conservation, preventing tourists from taking coral, building good sewage systems due to proximity to the sea, using solar heating systems for energy, and composting. For 77% of respondents, these individual actions toward environmental conservation came as a result of being involved in Dominican tourism. While government policies over time have stated the importance of nature as the tourism focus in Dominica, government actions have at times been contrary to this policy. Some authors have noted the Dominican government's ambivalence over promoting nature-based tourism versus mass tourism, and actions such as advocating attracting foreign investors, including large hotel chains, solidify this concern (Klak, 2007; Weaver, 2004). Both Weaver (2004) and Klak (2007) suggest that this wavering stems from the actions of a fiscally desperate government seeking economic development by any means possible. In Dominica, this conflict has frustrated many individuals and communities involved in tourism, and explains why so many respondents expressed the feeling that the government could do more to conserve the natural environment for tourism (Table 5).

Dominica is home to the Caribs (or Kalinago), the last remaining community of Indigenous people in the region. The Caribs, who provide a unique cultural attraction in Dominica, are located in the Carib Territory on about 3,700 acres of land on the north east coast of Dominica. Caribs were particularly hard hit with the decline in the banana industry since many Caribs were cultivators or banana plantation labourers. In 1993, the Carib Community formulated a management plan with government assistance to promote ecotourism in their Territory (HaySmith & Harvey, 1995). As a result, some Carib handicraft workers (Figure 6) are reporting income sales of $3,820 per year (Slinger, 2000). This amount was $1,000 more than the per capita gross national product of Dominica at that time (World Bank, 2000). Other economic opportunities have been created in the Carib Territory as a result of ecotourism, including employment and income earnings as tour guides, guest house owners, and local dance and cultural groups who perform for tourists. Nature sites within the Territory include hiking and biking trails, rivers, waterfalls and scenic ridges. An interesting hiking site is a lava formation called L'Escalier Tete-Chien (stairway of the snake) that emerges from the sea into the mountains. It is said that this stairway enabled the Caribs to visit the mystical snake to obtain special powers. With an increasing number of visitors who have an interest in botany and wildlife, Carib tour guides are expected to have knowledge of the flora and fauna in the Territory, and of Dominica in general

**Figure 6.** A Carib Woman with her Craft Items for Sale in the Carib Territory (Photo: taken by author).

In terms of cultural preservation, recent efforts have occurred to document the dying Carib language and other cultural information for the revival of crafts, medicines, and traditional music. The report of an examination by Boxill and Severin (2004) concluded that the social impacts of ecotourism on the Caribs are largely positive. A Carib Cultural Village was created as a visitor centre. It was, however, largely rejected by most Caribs who felt that outsiders had too much influence in its running and only benefitted a few particular Carib families (Sam Raphael, personal communication, 2008). Recent plans involving the Dominican government, an International Resort Company (Distinguished International), and the Carib Council, are to modify the Carib Cultural Village in order to develop an EcoLodge Resort and Conference Centre in 2009. The multi-phased projects will feature 100 villas, cottages, and tree-house suites, a 3000 square foot conference centre, dining facilities, spa services, and river-fed natural pools throughout the site. The resort design will incorporate solar heating and indigenous building materials (Hotel News Resource, 2008). The resort could potentially increase tourism earning opportunities in the Territory, one of the poorest communities on Dominica. However, like the original Carib Cultural Village, if it is perceived as being too influenced by outsiders or only for the benefit of a few particular Carib families, then it will face many challenges and be limited in its success.

## 3.4 Research Analysis

Interestingly, the model of ecotourism found in Dominica shows significant local benefit through ownership, management and general employment opportunities, as well as a potentially higher multiplier effect due to greater linkages between tourism and other sectors of the economy. So far, government policy in Dominica has been beneficial to small-scale locally owned facilities, and with the ecotourism's focus on natural features in rural locations, this type of development has occurred throughout the island.

This research suggests that certain factors have allowed Dominica to evolve into an ecotourism destination with high levels of local ownership and employment, significant backward linkages, and important regional development. Limited level land has made the construction of a major international airport prohibitively expensive. The runway length at Dominica's Melville Hall Airport in the northeast, 1585 meters, and Canefield Airport in the south, 792 meters, allow only small aircraft to land and neither presently has the facilities for night landings (Wilkinson, 1993, eSTANDARDSFORUM, 2009). Travel to Dominica generally requires a stop-over in such places as Puerto Rico or Jamaica, before arriving on the island. Dominica's physical features, its small size, its island geography, and the lack of an international airport may have acted as blessings in disguise, since they have been discouraging factors for both a high level of foreign visitation and foreign investment into larger tourism facilities. As a result, the facilities that have opened on the island are mostly locally owned, and many are small due to the limited financial resources available to most Dominicans. These are the type of facilities that the government's ecotourism policy has promoted. Because the focus has been on smaller scale establishments, many Dominicans may have felt capable of entering this industry since they would not have to compete with large, well-financed facilities.

The research also suggests that the small-scale, local nature of ecotourism on Dominica has contributed to a greater number of linkages between tourism and other sectors on the island. Larger facilities tend to demand economies of scale that are not available on Dominica and therefore encourage the importation of goods to supply the industry. With the exception of specific products that are not available on the island, the smaller tourism facilities on Dominica can, at present, be supplied by the local primary sector. This research showed high domestic ownership of facilities and a great desire to use locally made goods

Similar to other Latin American and African countries, government policies in Dominica, have tried to incorporate elements of environmental conservation into tourism planning. In other ecotourism destinations, this has sometimes meant that local populations have lost access to natural resources without obtaining alternative opportunities for their livelihood, particularly where foreigners dominate the ecotourism industry and its benefits. In some cases, this has caused locals to disregard environmental protection measures put in place by their government. In Dominica, individual conservation efforts of people involved in the ecotourism industry have been high due to a reliance on ecotourism and the environment

Finally, ecotourism seems to be benefiting the local indigenous community through greater recognition, revival, and maintenance of their culture, while providing them with a livelihood that is symbiotic with traditional activities.

## 3.5 Concerns

While this research has shown some definite benefits from ecotourism in Dominica, the situation is not perfect. Over time, several issues have caused concern over the direction of official national tourism planning on the island. Along with the already mentioned development of the cruise ship industry, other actions taken by the Dominica Government have led conservationists and Dominicans in the tourism industry to question their development motives. Controversial actions/decisions have included: (1) the lack of impact assessment at the additional tourist sites that are receiving increasing numbers of tourists (Watty, 2008); (2) the creation of hydroelectric power developments at Trafalgar Falls, Titou Gorge, and Freshwater Lake that have increased noise pollution in these tourist areas and diminished water flow to Trafalgar Falls, a key tourist attraction (Evans & Williams, 1997; Gayle, 1997); (3) consideration of a cable car system to take large numbers of tourists to the Morne Trois Pitons National Park. This was stopped in part due to public outcry and to the threat by UNESCO to strip the park of its designation as a World Heritage Site (UNESCO, 1998), (4) the way the Government voted at the International Whaling Commission conference in 2002, siding with Japan against a moratorium on whale hunting. Considering that the Dominican tourism industry also relies on whale watching to attract ecotourists, this vote seems contrary to action that would sustain this activity in Dominica (Black, 2005)5 ; (5) the plans of a previous Government to build an international airport that would involved displacing farming communities, destroying mountains and cutting down forest (Fontaine, 2003; Watty 2008)6 ; and (6) the push for a large hotel with foreign backing in the past. The 250 room, 400 time-share unit hotel was to be part of a controversial new 'economic citizenship policy', a government-initiated program whereby foreigners, particularly business people from the Pacific Rim, were sold a Dominican passport in exchange for cash (US$50,000) or for investing in certain (including tourism based) projects. Due to various reasons, this project was not completed (Nestmann, 2007).

Overall, the Dominican Government has put in place many policies that have encouraged the development of ecotourism on the island. While it is Government policy to encourage small-scale businesses and local participation, local funding and expertise is limited (Wilkinson, 2004). It remains to be seen if the Government's commitment to ecotourism as a main thrust for island development continues in the face of the constant desire for economic growth and pressure from stakeholders interested in the whaling, mineral, and forestry resources of the island. Furthermore, an unemployment rate of 23% and a net outward migration rate of 5.45/10007 population create a political and economic environment that pressures the government to grow faster in order to provide opportunities for young Dominicans who will otherwise seek them elsewhere (CIA World Factbook, 2009).

## 4. CONCLUSION

This study of Dominica shows that ecotourism can have the desired effects of promoting diverse regional development and increasing linkages with other economic sectors while encouraging more sustainable practices regarding the environment and positively impacting the local indigenous culture. The factors that helped achieve these favourable results are a combination of the natural limitations of this small island state along with a government focus on ecotourism development and the development of small scale locally owned facilities. The type of ecotourism model that has developed in Dominica may have transfer potential to similar, small island jurisdictions that are considering developing their tourism industries based on natural resources with the hope of stimulating significant linkages, maintaining local ownership, and facilitating more dispersed regional development.

The development of sun, sea, and sand tourism in the Caribbean has meant that many islands have been slow to promote their other environmental assets. Traditional tourism has typically been dominated by foreign ownership and expatriation of profits. For instance, over half the hotels in both Barbados and Jamaica were in foreign hands (Potter & Lloyd-Evans, 1997); while in Antigua, 35 out of the 37 major hotels were foreignowned (Pattullo, 1996). Potter & Lloyd-Evans (1997) have noted that up to 70 % of all tourist expenditure in the Caribbean leaves the islands and ends up in the hotel owners' foreign bank accounts. In contrast, the high percentage of local participation in the industry and the demand for local products suggests that the leakage factor of Dominican tourism is significantly less than the 70% recorded on other small islands by Milne (1990) and Potter and Lloyd-Evans (1997).

On many islands with traditional tourism, there also exists a spatial effect characterized by pockets of wealth surrounded by the general poverty of much of the local people. In Jamaica, 88% of the hotel rooms are found in only four locations along the North Coast (Price Waterhouse Coopers, 2004). These pockets of wealth attract locals in search of employment and a perceived chance of a better life; however, tourism employment opportunities for locals in Caribbean islands and other parts of the developing world have been criticized as being low-paying and servile in nature (Cukier, 1998). Pattullo (1996) gives the example of St. Lucia, where there was an estimated 90/10 split between expatriates and locals in senior management positions in the hospitality industry. In contrast, research for this article suggests that ecotourism on Dominica is creating opportunities for a high degree of local participation at all levels of the society in both urban and rural areas of the country.

The urgent need for economic diversification, dispersed regional development, and protection of local environments on other Caribbean islands has led to the search for development alternatives. The physical beauty and diverse flora and fauna of several Caribbean countries such as Grenada, St, Lucia, St. Vincent, Jamaica, and other small island states, offer opportunities for the development of ecotourism. The development of ecotourism in Dominica has significant relevance by providing a potential model for these other islands to follow.

## ACKNOWLEDGEMENTS

The author thanks the people of Dominica who assisted with the collection of data, interviews, logistics, and other information. The author also thanks Brian Friedman and anonymous reviewers for their feedback and editorial comments.

## REFERENCES

1. Archer, E. (1998) 'Tourism in Barbados' in D. Barker, C. Newby, & M. Morrissey (eds.) A Reader in Caribbean Geography, Kingston, Jamaica, Ian Randle Publishers, pp. 172-178.

2. Asiedu, A.B. (1997) 'Prospects for an Emerging Tourism Industry in Ghana', Research Review (new series), Vol. 13, Nos. 1 & 2, pp. 11-26.

3. Ayala, H. (1996) 'Resort Ecotourism: a paradigm for the 21st century', Cornell Hotel and Restaurant Administration Quarterly, Vol. 37, No. 5, pp. 46-53.

4. Black, R. (2005) 'Caribbean Call to Resume Whaling', BBC News, June 22, http://news.bbc.co.uk/2/hi/science/nature/4117888.stm

5. Blommestein, E. (1995) 'Sustainable Tourism in the Caribbean: An Enigma?' in M. Griffith & B. Persaud (eds.) Economic Policy and the Environment: The Caribbean Experience. University of the West Indies, Mona, Jamaica, Centre for Environment and Development, pp. 191-220.

6. Bonnerjea, L. & Weir, A. (1996) Commonwealth of Dominica Poverty Assessment: Report prepared for the Government of the Commonwealth of Dominica, Roseau, Dominica. Bridgetown, Barbados, British Development Division in the Caribbean.

7. Boxill, I. & Severin, F. (2004) 'An Exploratory Study of Tourism Development and Its Impact on the Caribs of Dominica', International Journal of Hospitality & Tourism Administration, Vol. 5, No. 1, pp. 1-27. Britton, S.G. (1982) 'The Political Economy of

Tourism in the Third World', Annals of Tourism Research, Vol. 9, No. 3, pp. 331-358.

8. Brohman, J. (1996) 'New Directions in Tourism for Third World Development', Annals of Tourism Research, Vol. 23, No. 1, pp. 48-70.

9. Caribbean Conservation Association (CCA) (1991) Dominica: Country Environmental Profile. St. Michael, Barbados, CCA and Island Resources Foundation.

10. Caribbean Net News (2009) 'Dominicans urged to continue increasing Country's Market Share', January 7, www.caribbeannetnews.com/dominica/dominica.php?news_id=13382&start=40&category _id=31

11. Caribbean Tourism Organization (CTO) (2004) 'Dominica', Division of Tourism, National Development Corporation, Department of Statistics, Dominica, www.onecaribbean.org/content/files/2004 dominica.pdf.

12. Caribbean Tourism Organization (CTO) (2007) 'Dominica', Division of Tourism, National Development Corporation, Department of Statistics, Dominica, www.onecaribbean.org/content/files/Dominica2006.pdf.

13. Carter, E. (2008) 'Dominica drops Support for Commercial Whaling', TheDominican.net, June 10, www.thedominican.net/2008/06/dominica-drops-support-for-commercial.html.

14. Central Statistical Office (CSO) (1999) Annual Travel Report. Roseau, Dominica, Government Statistical Office.

15. CHL (1999) 'Ecotourism Development Programme Design and Costing Study. Volume II Stabex Community Development Programme', Dublin, Ireland, CHL Consulting Group.

16. CHL (2006) 'Tourism Master Plan 2005 – 2015. Consulting services for the Preparation of a Tourism Master Plan 2005-2015. Commonwealth of Dominica, 2nd Draft Final Report' London, U.K., CHL Consulting Group.

17. CIA World Factbook (2000) 'Dominica', www.umsl.edu/services/govdocs/wofact2000/geos/do.html.

18. CIA World Factbook (2009) 'Dominica', www.cia.gov/ library/publications/the-worldfactbook/geos/do. html.

19. Caribbean Tourism Organization (CTO) (1993) Dominica Visitor Survey, ChristChurch, Barbados.

20. Cukier, J. (1998) 'Tourism Employment and Shifts in the Determination of Social Status in Bali' in G. Ringer (ed.) Destinations: Cultural Landscapes of Tourism, New York, Routledge, pp. 63-79.

21. Douglas, S. (2008) 'We are in the Home Stretch says Hon. Charles Savarin during Site Visit at Melville Hall Airport', Government of the Commonwealth of Dominica, press release, November 24, www.dominica.gov.dm/cms/index.php?q=node/674.

22. Duval, D.T. (2004) 'Trends and Circumstances in Caribbean Tourism' in D.T. Duval (ed.) Tourism in the Caribbean: Trends, Development, Prospects, London, Routledge, pp. 1-22.

23. Eco (2008) 'IWC Vote Buying', Eco International Whaling Commission Annual Meeting, June 2, Eco, Vol. 60, No. 1, www.asoc.org/Portals/0/2008%20Eco%20No%201.pdf.

24. Encyclopedia of the Nations (2008) 'Dominica: Agriculture', www.nationsencyclopedia.com/economies/Americas/Dominica-Agriculture.html.

25. eSTANDARDSFORUM (2009) 'Country Brief: Dominica January 20, 2009', eStandardsForum Financial Standards Foundation, www.estandardsforum.org/secure_content/country_profiles/cp_63.pdf.

26. Euromonitor International (2007) 'Travel and Tourism in Dominica', Research and Markets, www.researchandmarkets.com/reports/515169/travel_and_tourism_in_dominica.

27. Euromonitor International (2008) 'Travel and Tourism in Dominica', www.euromonitor.com/Travel_And_Tourism_ in_ Dominica.

28. Evans, P.G.H. & Williams, D. (1997) Development and Management of Nature Sites - Integrating Conservation with Ecotourism in Dominica, Project Report No. 4, Brussels, European Community Project No. B7-5040-24. Ecosystems Ltd.

29. Fontaine, T. (2003) 'No International Airport for Dominica', TheDominican.net, No. 42, May 7, www.thedominican.net/ articles/ airport.htm.

30. Gayle, D.J. (1997) 'Ecotourism: Fad or Future?', Hemisphere: A Magazine of the Americas, Vol. 8, No. 1, pp. 20-23.

31. Halcrow Group Ltd. (2003) 'Caribbean Development Bank Government of the Commonwealth of Dominica Country Poverty Assessment, Final Report Volume 1 of 2: Main Report', London, Halcrow Group Ltd.

32. HaySmith, L. & Harvey, J. (1995) Nature Conservation and Ecotourism in Central America. New York, Wildlife Conservation Society.

33. Honey, M. (1999) Ecotourism and Sustainable Development: Who Owns Paradise? Washington DC, Island Press.

34. Honey, M. (2008) Ecotourism and Sustainable Development: Who Owns Paradise? 2 nd edition, Washington DC, Island Press.

35. Honychurch, L. (1991) Dominica: Isle of Adventure, Basingstoke, Macmillan. Hotel News Resource (2008) Government of Dominica Supports Eco-Lodge Project Proposed by Distinguished International Hotels, Residences and Resorts, http://www.hotelonline.com/News/PR2008_3rd/Aug08_DominicaEcoLodge.html.

36. Hulme, P. (2007) 'Meditation on Yellow: Trade and Indigeneity in the Caribbean' in L. Dale & H. Gilbert (eds.) Economies of

Representation 1790-2000: Colonialism and Commerce, Aldershot, Ashgate, pp. 3-16.

37. Hypolite, E., Green, G.C. & Burley, J. (2002) 'Ecotourism: its potential role in forest resource conservation in the Commonwealth of Dominica, West Indies', International Forestry Review, Vol. 4, No. 4, pp. 298-306.

38. Johnson, D. (2006) 'Providing Ecotourism Excursions for Cruise Passengers', Journal of Sustainable Tourism, Vol. 14, No. 1, pp. 43-54.

39. Klak, T. (1998) 'Caribbean Development Policies in a Neoliberal Era: Case Studies' in T. Klak (ed.) Globalization and Neoliberalism: The Caribbean Context, Oxford, Rowman & Littlefield Publishers, pp. 111-134.

40. Klak, T. (2007) 'Sustainable Ecotourism Development in Central America and the Caribbean: Review of Debates and Conceptual Reformulation', Geography Compass, Vol. 1, No. 5, pp. 1037-1057.

41. Klak, T. & Flynn, R. (2008) 'Sustainable Development and Ecotourism: General Principles and Eastern Caribbean Case Study' in E. Jackiewicz & F.J. Bosco (eds.) Placing Latin America, Lanham MA, Rowman and Littlefield, pp. 115-136.

42. Latin American Bureau (1987) Green Gold: Bananas and Dependency in the Eastern Caribbean, Birmingham, Third World Publications.

43. McElroy, J.L. (2004) 'Global Perspectives in Caribbean Tourism' in D.T. Duval (ed.) Tourism in the Caribbean: Trends, Development, Prospects, London, Routledge, pp. 39- 56.

44. Ministry of Tourism and National Development Corporation (2005) Dominica: Tourism 2010 Policy. Roseau, Dominica. Milne, S. (1990) 'The Impact of Tourism Development in Small Pacific Island States: An Overview', New Zealand Journal of Geography, Vol. 89, pp.16-21.

45. Nestmann, M. (2007) 'Whither Dominica', http://nestmannblog.sovereignsociety.com/2007/04/whither_dominic.html.

46. Patterson, T. & Rodriguez, L. (2003) 'The Political Ecology of Tourism in the Commonwealth of Dominica' in S. Gössling (ed.) Tourism and Development in Tropical Islands: Political Ecology Perspectives, Cheltenham, Edward Elgar, pp. 60-87.

47. Pattullo, P. (1996) Last Resorts: The Cost of Tourism in the Caribbean, New York, Cassell.

48. Place, S.E. (1998) 'How Sustainable is Ecotourism in Costa Rica?' in M.C. Hall & A.A. Lew (eds.) Sustainable Tourism: A Geographical Perspective, New York, Longman, pp. 107-118.

49. Potter, R.B. & Lloyd-Evans, S. (1997) 'Sun, Fun & a Rum Deal: Perspectives on Development in the Commonwealth Caribbean', FOCUS, Vol. 44, No. 4, pp. 19-26.

50. Price Waterhouse Coopers (2004) 'The Jamaican Hotel Industry: 2004', www.pwc.com/cb/images/pdf/jamaicahotel2004.pdf.

51. Progressive Policy Institute (2004) 'Caribbean Banana Exports are Down', May 12, www.ppionline.org/ppi_ci.cfm? knlgAreaID=900003&contentID=252628.

52. Shamsub, H., Albrecht, W., & Dawkins, R. (2006) 'Relationship between Cruise-ship Tourism and Stayover Tourism: A Case Study of the Shift in the Cayman Islands' Tourism Strategy', Tourism Analysis, Vol. 11, pp. 95-104.

53. Slinger, V. (2000) 'Ecotourism in the Last Indigenous Caribbean Community', Annals of Tourism Research, Vol. 27, No. 2, pp. 520-523.

54. UNESCO (1998) Convention Concerning the Protection of the World Cultural and Natural Heritage, Bureau of the World Heritage Committee, 22nd Session, Paris, France, http://whc.unesco.org/archive/repbur98a.htm US Department of State (2009) Background

Note: Dominica, Under Secretary for Public Diplomacy and Public Affairs, www.state.gov/r/pa/ei/bgn/2295.htm.

55. Visit Dominica (2003-2009) 'Visit Dominica', www.visit-dominica.com/form2.cfm.

56. Watty, W.R.F. (2008) Feature Address W. R. Franklin Watty, Chairman of Diaspora Affairs, Dominica Academy of Arts and Sciences at the Annual Awards Gala Dominica Hotels and Tourism Association, http://da-academy.org/Watty_Address_DHTA.pdf.

57. Weaver, D.B. (1991) 'Alternative to Mass Tourism in Dominica', Annals of Tourism Research, Vol. 18, No. 2, pp. 414-432.

58. Weaver, D.B. (1998) Ecotourism in the Less Developed World, New York, CAB International. Weaver, D.B. (2004) 'Manifestations of Ecotourism in the Caribbean' in D. T. Duval (ed.) Tourism in the Caribbean: Trends, Development, Prospects, New York, Routledge, pp. 172-186.

59. Westbrook, T., Mendelssohn, G. & Evans, P.G.H. (1997) Stayover Visitor Survey 1995- 1996. Project Report No. 6: Integrating Conservation with Ecotourism in Dominica, Brussels, European Community Project No. B7-5040-24, Ecosystems Ltd.

60. Wilkinson, P.F. (1993) Tourism Policy and Planning in the Eastern Caribbean: Anguilla, Barbados, Dominica and St. Lucia - Report on a Research Project, Ontario, Canada, York University, Faculty of Environmental Studies.

61. Wilkinson, P.F. (2004) 'Caribbean Tourism Policy and Planning' in D.T. Duval (ed.) Tourism in the Caribbean: Trends, Development, Prospects, New York, Routledge, pp. 81- 98.

62. Woods, L.A., Perry, J.M. & Steagall, J.W. (1998) 'Tourism in Belize' in D. Barker, C. Newby & M. Morrissey (eds.) A Reader in Caribbean Geography, Kingston, Jamaica, Ian Randle Publishers, pp. 179-190.

63. World Bank (2000) World Development Indicators 2000, CD-ROM, Washington DC, World Bank.

64. WTO (2009) 'Regions: The Americas', UNWTO World Tourism Barometer, Vol. 6, No. 1, pp. 21-25, www.unwto.org/facts/eng/pdf/barometer/UNWTO_Barom08_1_en.pdf.

11

# 'GENUINE JERSEY': BRANDING AND AUTHENTICITY IN A SMALL ISLAND CULTURE

**Henry Johnson**

University of Otago New Zealand

## ABSTRACT

Jersey has attained a recognized international reputation especially in agriculture, tourism and finance. Over the past century, this small island has developed rapidly as a tourist destination and, since the 1960s, as a leading international finance centre. This paper discusses how a public-private organization uses a notion of islandness in order to help add value to local produce and products, and at the same time offering a sense of authenticity in terms of provenance. As an organization and brand, "Genuine Jersey" was launched in 2001 and is now a particularly visible island-based brand that does much to support local businesses and promote selected island produce and products more broadly to locals and visitors alike, as well as within a wider export industry. Drawing on discourses mainly from island studies and marketing, the article discusses how and why this brand exists on Jersey. While including a critical discussion of the brand itself, the paper shows how Genuine Jersey operates on and as a result of this particular island context.

## KEYWORDS

authenticity, branding, Genuine Jersey, island, Jersey, produce, products, publicprivate partnership

# 1. INTRODUCTION

This paper studies the interconnection between branding and constructions of authenticity on Jersey. The aim of the article is to show how one organization on a small island utilizes a notion of islandness in several ways in order to help add value in terms of provenance to select local produce and products. The organization functions in this context through an interconnection with government and business in a way that is constructed as a result of its specific island context. It is here that this paper provides a critical discussion of not only how such branding occurs in this island setting, but also how the island itself adds value when positioning local produce and products.

As a relatively small island with a land area of about 118 km2 (States of Jersey Statistics Unit, 2010: iii) and a population in 2011 of 97,857 (States of Jersey Statistics Unit, 2012: 2), Jersey boomed as a tourist destination in the decades following the Second World War, especially with visitors from the United Kingdom who were attracted by such features as its warmer climate, low taxes and natural tourist attractions. Jersey is approximately 137 km from England and 23 km from France (States of Jersey Statistics Unit, 2010: iii). However, the mass tourism the island once experienced declined rapidly in the 1990s under pressure from other tourist destinations as a result of increased and cheaper travel options. In its place, and ever growing from the 1960s, finance has now become Jersey's main industry. For a small island, the physical landscape has subsequently been altered in that many hotels that once housed decades of visiting tourists have now been replaced by apartment blocks for the growing permanent and temporary workers that dominate the island's business sector (Lichrou et al., 2010: 139).

Jersey has its own government (the States of Jersey), laws, taxes, endangered language and various other aspects of governance and cultural heritage that contribute to making the island what it is today. Each helps form the Jersey brand. Nevertheless, even though some features of Jersey's heritage are foregrounded in tourism and other contexts of island celebration, such as British and German military defences (the island was occupied by Nazi Germany from 1940 to 1945), it is an identity that is actually constructed from a diverse population and increased external influences. For example, based on figures from the last official census in 2011, the place of birth of Jersey's population showed mixed influences that were the result of recent migrations from the British Isles, the European mainland and further afield (Table 1).

**Table 1.** Islanders' Place of Birth (States of Jersey Statistics Unit, 2012: 9).

| Place of Birth | Number of People | Percentage of Population |
|---|---|---|
| Jersey | 48,653 | 50 |
| British Isles | 30,223 | 31 |
| Portugal/Madeira | 7,031 | 7 |
| Poland | 3,133 | 3 |
| Republic of Ireland | 1,880 | 2 |
| Other European country | 3,146 | 3 |
| Elsewhere in the world | 3,791 | 4 |

In the broader setting of the European Union, of which Jersey is not a part (nor is it a part of the United Kingdom),1 as well as other outside influences, Jersey today makes much of its uniqueness as an island (e.g. location, size, history, politics) to celebrate its heritage and selfgoverning identity, particularly in the spheres of agriculture, tourism and finance. It is in this context that branding and constructions of authenticity interconnect to celebrate products that stand for Jersey. Selected goods, whether manufactured, grown or caught, are presented by Genuine Jersey (a public-private organization; also known as Genuine Jersey Products Association [GJPA] in full) as types of "cultural heritage assets" (Ivanovic, 2008: 215), and these items are given value and consumed by locals and visitors alike in a process that not only supports some local businesses, but also actively celebrates products that are considered unique to the island, whether well or recently established.

In connection with understanding these links, this paper discusses the Genuine Jersey brand, which was launched in 2001, as a case study for investigating critically some of the ways this small island culture celebrates and constructs notions of authenticity in its local economy. As a way of contributing to and extending the literature on the branding of islands (e.g. Baldacchino, 2005, 2010b; Grydehøj, 2008, 2010; Kelman, 2007; Khamis, 2010, 2011;

Leseure, 2010; Lichrou, O'Malley & Patterson, 2010; Pounder, 2010; and Reddy & Singh, 2010), this paper offers a study of a public-private branding exercise of one organization that is based around the notion of the authenticity of the goods it supports and promotes. The activity is seen to be inherently based around place branding, goods branding and the construction of a notion of authenticity. The discussion raises questions in connection with the selection criteria of Genuine Jersey products, as well as pointing out contradictions inherent in the branding process itself. Why has a notion of authenticity been constructed? How can a public-private organization truly represent the wider island community? Who are the key players in this branding process? In this context, Genuine Jersey is seen to undertake a process of selecting certain Jersey produce and products to brand as genuine island goods, an activity that helps show characteristics of a commercial exercise that actually interconnects business and government. In doing this, the paper also provides an example of how a small island with a recent history enmeshed in the tourist industry is reinventing itself in an age of rapid change to the three main industries that have themselves often come to stand for the island: agriculture, tourism and finance (Table 2).

**Table 2.** Principal Industries in 2009 (States of Jersey, Jersey Facts and Figures).

| Principal industries (2009 figures) | % of Gross Value Added |
|---|---|
| Financial services | 43 |
| Public administration | 8 |
| Wholesale and retail | 7 |
| Construction | 6 |
| Tourism | 3 |
| Agriculture | 2 |
| Other business activities | 9 |

This article draws broadly on case studies of and critical discourse on place branding from several scholarly fields, especially island studies (e.g. Baldacchino, 2010c; Grydehøj, 2008; 2010) and marketing (e.g. Askegaard & Kjeldgaard, 2007; Manniche et al., 2009; Manniche & Larsen, 2009). The article provides a case study of place and produce/product branding on Jersey. As Kavaratzis (2008: 51) has noted:

Places have long felt a need to differentiate themselves from each other, to assert their individuality in pursuit of various economic, political or socio-psychological objectives. The conscious attempt of governments to shape a specifically designed place identity and promote it to identified markets, whether external or internal, is almost as old as civic government itself.

In this setting, and building on Kavaratzis' points, this article shows that there are several layers of branding in the Jersey case, including island/place branding and product/produce branding. Each layer contributes to creating not only a public-private organization that brands a place and its goods as authentic, but also a branding proposition with a sense of islandness and 'othering' at its core.

Several research methods have been used for data collection and analysis. Secondary data have been important for assembling background information and gaining historical knowledge of key facts and figures, as well as comparing different types of case study approaches to similar topics (e.g. Khamis, 2010; 2011; Reddy & Singh, 2010; Pounder, 2010; Witkowski & Jones, 2006). The research also included a telephone interview in May 2011 with the Chief Executive Officer of Genuine Jersey. Undertaking one in-depth semi-structured key-informant interview was essential in this particular context (the interviewee is the person responsible for the day-today operations, and the subject of this article is on this organization). Such an approach is typical in the field of cultural biography (e.g. Parke 2002), although in the context of this article, rather than focussing on the person, emphasis was placed on the organization. In other words, this is a case study of one organization and relies on several sources of information: historical, verbal and observational. The interview was based on open-ended questions on such themes as background, organization, funding, purpose and activities. Consent was given to use names and quotes from the interview, on the basis that everything said was typical of a media interview that would be broadcast to the wider public.

Following a short section that contextualizes recent place branding in connection with Jersey from a politico-economic perspective, the main part of the discussion focuses on outlining and discussing the Genuine Jersey brand: its purpose, structure, logo and activities.

## 2. BRANDING JERSEY FOR THE TOURIST GAZE

Many islands have utilized, rediscovered or even invented aspects of their heritage as part of a branding exercise. Grydehøj (2010), for example, explores the idea of heritage production in Shetland (Scotland), Åland (Finland) and Svalbard (Norway). In earlier research, and in connection with the influence of business, he points out that:

It is not just that marketers use marketing to disseminate a message about a product. The complex and evolving nature of the product (the place) itself and the marketers' frequent personal identification of themselves as part of the product means that the very process of place branding has the dual potential to inform policy makers and to intensify their prejudices (Grydehøj, 2008: 175).

Such research, as well as that mentioned in the introduction, shows that some small islands have made much of their environmental or cultural settings in terms of creating a place brand for the island, or for the promotion of local industry, including tourism and commercial goods.

The Jersey brand, however, has several guises, and has over the past sixty years or so been adapted in various ways. In terms of branding itself as a location (an island location), Jersey's main industries have been at the heart of each successive branding and re-branding exercise. While agriculture, tourism and finance have distinct historical and recent histories on the island, their presence and importance have impacted at different times and in different ways in connection with Jersey and place branding. Although a sixty-year period is a relatively short time span to consider, the changes that have occurred to Jersey's principal industries have been dramatic, and the decline of some (in place of the growth of others) is currently seen at a juncture where past successful industries have acquired a status akin to cultural heritage (Ivanovic, 2008: 215). That is, while their importance to the island's economy may have been transformed from principal to secondary status, they have maintained importance to a smaller yet still significant market, and at the same time attained status in terms of their heritage on Jersey as part of the Jersey brand in its broader guise.

As the most southerly inhabited island in the geographic British Isles (using a political definition of the term),2 Jersey boasts a milder and sunnier climate to the UK, and as such has been a traditional source for early agricultural crops and especially unique local produce such as the Jersey Royal Potato. Indeed, parts of the island itself have physically been re-defined as a result of centuries of farming, whether crops or dairy with the well-known Jersey cow. As the States of Jersey (i.e. Jersey's government) branding guide notes, "farming has shaped the beauty of Jersey's landscape and sustained local families over generations" (States of Jersey, 2007: n.p.; see also Corporate Edge Branding, 2008). This particular branding exercise was based around several questions relating to the promotion of Jersey: "How can we position Jersey as a sophisticated and contemporary place to visit? How can we reinforce perceptions of strength, solidity and integrity within the financial market? How can we help grow the agricultural sector in recognizing (and marketing) the unique qualities that make Jersey produce worth trying and buying? How can we reawaken a sense of pride amongst the whole community in Jersey?" (States of Jersey, 2007: n.p.). In this context of top-down branding, agriculture is foregrounded, along with finance, tourism and community, as one of the overarching features that the States of Jersey aims to promote in connection with standing for Jersey in a broader sense (other spheres of this branding exercise are discussed later). It is agricultural produce like this that are typically supported by Genuine Jersey, although the breadth of produce promoted by this public-private organization moves far beyond such traditional crops as the Jersey Royal Potato.

The growth of the post-Second World War tourist industry was primarily the result of an increased ability to travel during economic growth and increased amounts of leisure time. During this period, Jersey put itself forward as being sunny, close to France and having a social and cultural heritage that was unique amongst the British Isles (such as Norman influence, beaches around much of its coastline, low or no taxes, war artefacts), which attracted mass tourism from the UK.3 In this context, Jersey has histories that are well-known outside the island, reputations that allow some aspects of its past and present to be foregrounded and to stand for the island.

Jersey's beaches, of which the small island boasts many, as well as its physical landscape – with small lanes, walking paths and sea views – offer the tourist industry various facets that have helped attract many tourists over the years. As noted in the recent corporate branding exercise:

Jersey has always been a very beautiful, very desirable place to live and work. And it already has a 'brand': a set of perceptions that many people share about what Jersey is like. Safe, secure, beautiful, great beaches, a little old fashioned, off shore, Jersey Royals, wealthy, a bit inward looking. A whole mix of the good and bad, the fair and the unfair (States of Jersey, 2007: n.p.).

In this setting, the heyday of Jersey's mass tourism that dominated the island until around the 1990s witnessed a multitude of hotels and guest houses shaping the built-up parts of the island, with large hotel complexes surrounding several beaches, tour companies filling the roads with coaches, and the ever-present hire-car (Brychan & Thomas 2012). Nevertheless:

Traditional Jersey industries such as agriculture and tourism were superseded by financial services as the dominant industry in Jersey. The financial services sector (which includes banking, trust and company administration, fund management, accountancy and legal activities) has grown such that it now accounts for almost half of total economic activity in Jersey and employs about a quarter of the workforce (States of Jersey Statistics Unit, 2010: iv).

Branding on Jersey, therefore, is undertaken in many ways. As a place, Jersey is known both for its individual industry sectors as noted above, and collectively for each of them as a way of representing the island. However, as discussed next, when place branding and goods' branding are combined in a public-private partnership, a different type of marketing dynamic is revealed with regard to the produce and products that form part of that brand vis-à-vis those that are not. Moreover, those goods that are accepted as part of such a brand might be viewed as part of a staging process or emergent authenticity that actually binds them to inauthentic experiences (Cohen, 1988: 373; MacCannell, 1999).

# 3. GENUINE JERSEY AND THE AUTHENTIC ISLAND

## 3.1 Purpose

Jersey holds an enviable position amongst other small island jurisdictions and cultures in terms of the international recognition of several of its products and produce. Household and even global brands include Jersey milk (from the Jersey cow), the Jersey Royal Potato, the Jersey tomato, Jersey cabbage and broccoli. As discussed later in connection with place branding, most of these products include the word "Jersey" in their name, something that helps brand not only the items but also the island (as with Shetland wool and Orkney beef). Furthermore, these products are well known within and beyond their home island, and their long association with Jersey would have already created a reputation as part of an established island brand (Baldacchino, 2005: 29). Before finance became the dominant industry on the island, and also part of Jersey's contemporary brand, local produce such as those listed above helped form an important local and export industry. For much agricultural produce, Jersey's warmer climate in comparison to the other British Isles helped local traders ship their produce to the UK at the start of the season. While Jersey produce such as these remain in high demand, their place on the island and in the export industry has taken a slightly different role. That is, while they no longer form part of the main industry for most islanders, they still hold an important place as products for local and non-local consumers, as well as for their ability to stand for the island as part of its traditional past. Here, the island's heritage has been partly constructed around the reputation of its agricultural produce, and the preservation, and sometimes rediscovery, of some produce and products is found within the Genuine Jersey framework (Lichrou et al., 2010: 141).

Top-down branding exercises, such as the ones discussed by Grydehøj (2008: 175) in connection with Shetland and other islands, explore issues of generic island branding imposed on a "locally-oriented identity concept". In this setting, place branding serves to collectively give an island distinct characteristics that might help market the island to incoming tourists or locals or non-local consumers (Baldacchino, 2010b). The focus of this discussion, however, is on a Jersey organization whose sole purpose is to brand select local produce and products with a distinct label, "Genuine Jersey". As a non-profit marketing organization that is heavily supported by government (discussed below), Genuine Jersey works directly with many local businesses in order to help bring aspects of some island industries together to collectively brand unique features of Jersey (Genuine Jersey website). Genuine Jersey collaborates with such businesses by offering marketing advice and support, promoting members' brands by holding public events around the island, holding growing competitions such as the Genuine Jersey Royal Potato Growing Competition,4 and offering a branding logo (Figure 1) that members use when packaging, labeling and marketing their goods. In such a framework, Genuine Jersey offers a level of branding for local products that fall within its guidelines (discussed later). Such products might have a long history on the island, or they might be recently conceived in a broader commercial environment that is aligning itself with a notion of a Jersey brand as a way of achieving greater commercial advantage. As noted in more detail later, there are also products that may have an established place on the island, yet fail to be recognized by Genuine Jersey.

## 3.2 Structure

The organizational aspects of Genuine Jersey provide various links with Jersey's government, a government department and with local businesses, and a study of these interconnections helps show how the organization is able to exist in a public-private sphere and in connection with Jersey's small island political and commercial landscape. This public-private partnership allows government to collaborate through a series of organizations with local businesses, and for those businesses that are members of Genuine Jersey to indirectly draw on funding support that ultimately serves to help market their goods.5 Public funding through the States of Jersey Economic Development Department (Jersey Tourism is a part of this Department) is used to help some businesses promote their produce and products collectively under a shared brand that is inherently defined as local and authentic versus goods that are either imported, locallyproduced but do not meet the requirements of Genuine Jersey, or have not yet been offered to Genuine Jersey for consideration. Genuine Jersey's link with Jersey Tourism is very close. They share the same physical office space, and both are linked to the Department of Economic Development. It is this construction of a perceived island-centred authenticity that this discussion is mostly concerned with.

**Figure 1.** 'Genuine Jersey' Logo (Genuine Jersey, Branding Guidelines).

Launched in 2001 by several local businessmen "who wanted to be able to differentiate their product, which was locally produced, from product that was imported, but sometimes gave the impression that it was local" (Garton, 2011), Genuine Jersey Products Association (GJPA) interconnects a number of commercial and political affiliations, and Genuine Jersey is the brand name that is used in the public domain. Above this association there is now a company called Jersey Product

Promotion Limited (JPPL), which was formed in 2008. At this time, GJPA became a subsidiary of JPPL, the latter of which is "wholly owned by a special purpose trust under the aegis of the States of Jersey Economic Development Department" (Genuine Jersey, Branding Guidelines: 1).6 One of JPPL's primary purposes is to employ Genuine Jersey's Chief Executive Officer, and also to take more control of the activities of GJPA. The direct connection with government is that JPPL receives an annual grant from the Economic Development Department, and this Department has outlined some strict criteria when awarding the grant. As noted by its Executive Director, who is also Chief Executive Officer of Genuine Jersey, JPPL is essentially to "promote local product both in the local market place and overseas" (States of Jersey, 2010a: 11; see also States of Jersey, 2010b). In 2010, for example, JPPL received £140,000 (about US$217,000) from the Economic Development Department, half of which went to Genuine Jersey (States of Jersey, 2010a: 10). As well as this grant, Genuine Jersey receives office space and office equipment and fittings. When the differences between the various affiliations of Genuine Jersey are defined, in general terms, JPPL focuses on local product off island, and Genuine Jersey on local product on island, the latter being the focus of this discussion (States of Jersey, 2010a: 5).7

The Genuine Jersey Management Committee includes representation from government, government organizations, Genuine Jersey and local businesses. Its structure comprises the following positions: Chairman (independent), Vice-Chairman (independent), Chief Executive Officer, Committee Members (three from the scheme's membership and three from the Economic Development Department [one must be a Trading Standards Officer]), Associate from the Jersey Farmers' Union, and an Invited Arts and Crafts Representative. This structure and membership allows Genuine Jersey to be represented not only from government but also from key partnership organizations and business. The Chief Executive Officer, who is Executive Director of JPPL, is remunerated in this position and has the main day-to-day responsibility of the running of Genuine Jersey.

There are strict criteria for becoming a member and for using the Genuine Jersey branding logo once an annual payment has been made, as noted in its Charter:

2.6: In general, goods will qualify as being appropriate to carry the Genuine Jersey logo if they:

2.6.1: comply with all relevant legislation and the current assessment criteria applied by the Management Committee, 2.6.2: are designed and produced entirely in Jersey using exclusively local raw materials and skills, 2.6.3: are grown in Jersey from seed or acceptable imported stock 2.6.4: are designed and produced in Jersey not necessarily of local raw materials but where there is evidence of a level of skill and design showing the individuality of the designer/maker, 2.6.5: are produced in Jersey where a local component is the reason for the product being produced and using local skills in its production, 2.6.6: are produced, developed, grown or reared in Jersey from imported stock but where local skill, husbandry or some other distinctive element unique to Jersey plays the predominant role in production, 2.6.7: restaurants may apply for membership where they have dishes on the menu which contain at least 80% local ingredients. Such dishes only may be identified on the menu with a Genuine Jersey logo (Genuine Jersey, Charter).8

To become a member of Genuine Jersey, businesses must follow these criteria and pay an annual membership fee, and in doing so they are able to brand their goods using the Genuine Jersey logo. The emphasis on the goods is on the location of Jersey. That is, it is on the produce or product in terms of where it is grown, caught or made. The exception is in restaurants, where the dishes can have 20% non-local ingredients and still be eligible to join the organization. By varying the amount of non-local ingredients acceptable to become a member of Genuine Jersey, the organization is able to acquire further income from some of the numerous restaurants that are in Jersey, many of which service the tourist industry and the affluent culture that has emerged as a consequence of the finance industry. However, by reducing the criteria for the amount of non-local ingredients, questions are raised about broader aspects of membership. Should the membership criteria be the same for all businesses? Why should restaurants get such a reduction? What about other businesses that can only exist with a similar amount of imported items? In the modern-day globalized world, many businesses that may have roots in an island culture might rely in varying degrees on imported items. It is here that one can see the need for the organization to gain more members from representative spheres of island business, although in doing so the core of the "genuine" message is perhaps diminished when criteria are allowed to vary between businesses.

In terms of the authenticity of the goods that Genuine Jersey promotes, the organization clearly emphasizes the place of the island rather than the people producing the goods, although they should reside on Jersey. In doing so, a contradiction is revealed regarding the social place of production rather than the local place of production. For instance, the Jersey Royal Potato has been a traditional part of Jersey agriculture for many generations, but it is not always locals with many generations of Jersey ancestry that have been responsible for providing all the labour connected with farming this product. Jersey has a long history of importing short- and long-term agricultural migrant labour, as well as similar labour opportunities in other industries such as tourism and finance. The agricultural industry, for example, is known for its imported labour, first in terms of French workers, then Portuguese and more recently Polish. Likewise, the tourism industry, with its history of establishing a plethora of hotels and restaurants, has employed many migrant workers who have each contributed to the promotion of Jersey cuisine for local and tourist consumers. Other business too rely on Jersey's emerging multicultural labour force in the production of goods, each of which points to the emphasis on place in the promotion of Genuine Jersey goods rather than the people involved in their production. While the area of authenticity is also discussed later, comparisons can be made with the branding exercises of other islands. For example, Khamis (2011: 1) challenges the branding of the island of Lundy in the Bristol Channel in the UK as unspoiled or untouched: "this perception is highly misplaced, and fails to take stock of the considerable effort that goes into managing Lundy". In other words, Genuine Jersey seems to promote produce and products as part of a perceived authentic Jersey, something that represents the island in a kind of pure form (albeit with allowances made for some businesses). However, while Jersey in the present day has roots in its island heritage, the island is also part of a contemporary world of global flows with rapid movement of people, goods and information. Genuine Jersey may represent a genuine Jersey, but does it truly represent the genuine Jersey?

The mantra of buying local allows Genuine Jersey to align itself with contemporary green politics. This is especially evident in public statements by Genuine Jersey where the importance of purchasing its members' products is placed in the context of potential environmental disasters:

It gives consumers the choice to mitigate the adverse affects of global warming by reducing their food miles and the Island's carbon footprint. Making an informed choice to buy local and embrace seasonality is good for the Island's economy as it supports local jobs and recycles the local pound. Keeping agricultural fields in production also ensures that the Island's countryside remains green (Garton, 2010: 2).

The use of such a collective and public-private method of branding in Jersey reflects wider global trends that have seen the increased labelling of produce and products in terms of locality. The example suggested by Scherrer et al. (2009) of "Designation of Origin" labelling in the Canary Islands' wine industry and its connection with tourism is one example from an archipelago. The European Union has done much to try to ensure that goods with a recognized connection to a place are safeguarded against inauthentic use of the same name. This is "a status that recognizes these products to be authentic, quality, local products ... that are available to both locals and outside visitors looking for typical Canary Islands' foods" (Scherrer et al., 2009: 452; Sainz, 2002). Most importantly, in the same way that Genuine Jersey links island products and produce with the island itself, "Designation of Origin" operates by "highlighting originality and providing an authentic experience that links the local products with aspects of the landscape, the area's history and community in a holistic way" (Scherrer et al., 2009: 460). This type of branding might be viewed as protectionism where members' goods are given special status over other goods that attempt to link to the success of the original produce or product. For example, within the European Union there is a Protected Geographical Status (PGS) law that protects the naming rights of regional foods. For Genuine Jersey, however, the organization is collectively branding select island produce and products to show that they are genuinely local, rather than imported. This relates closely to the PGS law, although it does not exclusively lay claim to the name Jersey, but, rather, the location and manner in which the goods are produced.

The tropes of providing "pride for residents" and a "desire for visitors" (Genuine Jersey, Charter: 2) point to the provision of a brand that seeks to embody a sense of place in the goods it promotes. As noted by the Chief Executive Officer, Genuine Jersey is "trying to drive some level of quality into the product" (Garton, 2011). For the brand to succeed in the local market, islanders will need to take pride in it in comparison to imported or non-Genuine Jersey goods, and visitors will need to seek the brand as a memento of islandness during their short-term stay on Jersey (Figure 2). As noted by Jersey Tourism, "work continues to take place through the auspices of Genuine Jersey to encourage local producers to trade and to develop products so that visitors can take home a Genuine Jersey souvenir" (Jersey Tourism 2008: 10). It is here that Genuine Jersey points to the celebration of local and of members' goods vis-à-vis other goods. In such a context, membership of the organization becomes one of aligning a business and its goods with a brand that stands for Jersey, and only market research would be able to tell the success or lack of it in terms of whether it is beneficial for a business to become a member or not. There are indeed various businesses on Jersey whose goods are not recognized by Genuine Jersey, yet they claim to have very similar objectives. Take, for example, The Jersey Cow Company, which notes:

We endeavour to produce products of the highest quality at a price that makes them an 'everyday luxury indulgence'. The key to all of our products is 'purity & provenance'. We will only create products that have a genuine link to the island and that are not only good for you but also good for the environment (The Jersey Cow Company).

**Figure 2.** 'Genuine Jersey' Banners at Jersey Airport. (Photo by author.)

This company, as one example, has aims that are almost identical with those of Genuine Jersey, yet it is not a current member and thus is establishing its own brand. In such a context, some brands might actually be disassociating themselves with collective brands of the publicprivate type. Genuine Jersey has now been in existence for over ten years, and it has increased its membership on an annual basis. It is forming a brand that will suit the products and produce of some Jersey businesses and not others. In forging its overall membership, the organization is creating a division on the island regarding businesses that have an affiliation with Genuine Jersey and those that do not. Some businesses are not members either because they do not fulfill the selection criteria, or because they have chosen not to become affiliated with the organization. While such a topic is the study for future research, it will suffice to say here that the only main incentive in staying away from this brand, even if fulfilling membership criteria, would be in terms of being able to offer an alternative brand that has the ability to function in its own right and not associate itself with a collective based on authenticity.

Apart from government funding, Genuine Jersey receives income in the form of an annual fee from members and sponsors. In 2011, there were 91 members and seven sponsors (see Genuine Jersey, Our Members). Over the ten years of its existence (at the time of writing this article), Genuine Jersey has grown rapidly: 20 members in 2001; 60 members in 2005; and 91 members in 2011. While members' goods must qualify to gain labeling rights, in a somewhat contradictory way to the underpinning purpose of Genuine Jersey, sponsors simply pay an annual fee based on the size of the business, or a service for Genuine Jersey's members, and do not have to qualify first. In addition to the membership fee, the sponsors' scheme provides not only a further income stream for Genuine Jersey, but also further connections with the island's businesses. A reward from Genuine Jersey to its sponsors is the right to use the Genuine Jersey Sponsor mark, even though the business might not actually qualify to be a member (Genuine Jersey, Sponsors). There is obviously inconsistency of purpose in connection with products that comply with the Genuine Jersey membership criteria and sponsors who do not, yet the latter are still able to associate themselves with a brand that has authenticity of product at its core. This point of contention highlights how on the one hand a public-private partnership is able to function within the commercial market, yet on the other hand how commercialism can help undermine some of the core values underpinning the organization.

The point of difference with the Genuine Jersey brand is that it is simultaneously top-down and bottom-up in terms of its purpose, and is also a public-private partnership. While closely linked with a government department, on which it relies for considerable funding and infrastructural support, Genuine Jersey has a membership base of local companies and businesses that collectively identify with and use the Genuine Jersey logo and brand as a way of marketing their produce and products. As part of the funding agreement, the grant that Genuine Jersey receives is reduced each year, and is intended to be replaced by private funding. A consequence of this is that the number of sponsors is likely to increase, with the possibility that some may not fulfil the criteria for membership and therefore reducing the organization's overall association with businesses that reflect its primary purpose.

With such an organization, questions are raised when considering the incentives for local businesses to become members. What is in it for them? What if they don't join? As noted in its Charter, "the Genuine Jersey brand seeks to promote":

- Members' goods
- An informed choice for the consumer
- The retention of consumer spending in the local economy
- A pride for residents in the quality and diversity of the goods of their Island
- Visitors' desire to prolong memories of a Jersey holiday by buying Genuine Jersey goods
- An accord between participating businesses to enhance the brand and generate viable business for each other, without limiting consumer choice or imposing any price configuration on local goods
- Sustainability for and innovation amongst local producers
- A guarantee of compliance with all statutory requirements (Genuine Jersey, Charter: 2).

The Charter includes a number of key points intended for locals and visitors alike. While the underpinning purpose of the branding is surely to increase sales of members' goods as competition grows in local and global markets, and hence ensure the sustainability of the local economy with "the retention of consumer spending" (Genuine Jersey, Charter: 2), several of the Charter's points help show how the brand can embody meaning in other ways for consumers who purchase its goods. For example, by offering an "informed choice for the consumer" (Genuine Jersey, Charter: 2), Genuine Jersey goods are immediately juxtaposed with others that do not include the brand mark. This type of branding shows a labeling that embodies a notion of authenticity at its core. That is, Genuine Jersey goods represent a "genuine" Jersey, others, therefore, do not. But such a constructed type of authenticity of place also serves to render the brand inauthentic because the image is created, or at least reinforced, by the brand in the first place.

## 3.3 Logo

The purpose of any branding exercise, whether initiated by a government organization or a private company, is to offer a product with which people might easily identify and recognize. As one researcher has commented:

Today it is argued that not only should products and services develop a system of brand management focused on their identity, which helps develop a coherent execution, but places should also develop a similar brand management system (Freire, 2009: 420).

Jersey Tourism, for example, would brand itself and the island of Jersey for slightly different purposes than a local retailer, although there might be similarities in terms of the desired targeted market. A prominent feature of most contemporary branding exercises is the inclusion of a logo, which acts as a visual signifier that is usually easily understood and remembered (Klein, 2000). For example, Jersey Tourism has recently introduced its "golden bird" logo (see top right of Jersey Black Butter poster in Figure 3), which is used as an emblem for the island in terms of its recognition as a tourist destination (States of Jersey, 2007: n.p.). The "golden bird", which has also been described by unconvinced locals as a "flying banana" (e.g. Lewis, 2007: n.p.), was a £250,000 exercise (Quérée, 2007: n.p.), and is intended to represent "Jersey's sense of independent spirit and freedom, the richness of its culture and all its produce, and its fresh outlook for the future" (States of Jersey, 2007: n.p.).

Genuine Jersey too offers a logo as well as several slogans that inherently index a sense of authenticity in the language that is used.9 With the requirements of members' produce and products having to fulfil strict criteria in terms of showing how they are locally produced, the notion of authenticity is both inscribed in the goods themselves and also ascribed to them. The single aspect of branding these goods involves becoming a member of Genuine Jersey and being privileged to use its logo and branding network. Genuine Jersey provides extensive guidelines for the use of its logo and in its Charter notes that it is:

- an endorsement of the goods' own brand or trade mark,
- awarded only to those goods which fulfil both the spirit and the word of the Genuine Jersey Charter,
- available in various regulated colours to achieve compatibility with goods,
- used only in accordance with strict guidelines which are subject to regular review (Genuine Jersey, Charter: 2).

**Figure 3.** Shop at Jersey Airport. (Photo by author.)

Genuine Jersey might be seen to justify its existence by establishing itself against inauthentic brands being sold on Jersey that give the outward impression that they are a product of local traders, but in reality have been produced in part or entirely outside the island. One example of this that was offered to me during research was that of Jersey fudge being sold at Jersey Airport, a context ripe with touristic merchandise and a place where the visitor is able to purchase a lasting memento of the island. One particular brand gave the outward appearance of representing the island and was labeled "Jersey Fudge", yet it was produced in China (Garton, 2011). Another brand, however, Sue's Fudge, offers the consumer "real Jersey fudge" and is recognized by Genuine Jersey.

Authenticity (i.e. 'genuineness') for Genuine Jersey is about speaking what it perceives as the truth about local produce and products; Genuine Jersey makes a statement about the island and for its members' goods. In this context, "genuine" stands for Jersey, its provenance, and the brand has the infrastructure and governmental and establishment links to ensure that it makes an impact on the island for local businesses. Here, a sense of nationalism and localism (as well as island identity) is embodied in the Genuine Jersey logo, where "authenticity ... is constructed by cultural representation and media techniques, which makes it an inter-subjective discourse but therefore no illusion" (Fornäs, 1995: 219). It is in this context that "image makers and the media assume a more powerful role in the shaping of political identities" (Harvey, 1990: 289). By grouping different goods under one umbrella brand, or rather allowing different businesses to use its logo after payment to become members of the organization, Genuine Jersey is constructing a sense of "object-related authenticity", the "cognitive experience of the authenticity" (Ivanovic, 2008: 322). While "authenticity is a socially constructed concept" (Ivanovic, 2008: 323), the branding process would aim to achieve a higher level of desire for the goods being branded (Wang, 1999). Ironically, of course, such staged authenticity would actually be "damned to inauthenticity" (Cohen, 1988: 373; MacCannell, 1999). Furthermore, the Genuine Jersey brand groups a number of disparate goods under one label, which allows some goods to attain the status of authentic and to be associated with produce and products that have a longer recognized history on the island, such as the Jersey Royal Potato, cow or tomato (Baldacchino, 2005: 29-31).

To make the point even stronger, in its advertising campaigns Genuine Jersey asks the consumer to "Look for the logo before you buy" (Advertising flyer), and "Look for the Mark before you buy" (Advertising flyer). The same flyer offers six reasons why the consumer should buy local Genuine Jersey produce, which are:

- Eat seasonally
- Support local agriculture

- Reduce food miles
- Preserve the environment
- Support the local economy
- Provenance/traceability (Advertising flyer).

The language of authenticity in the term "genuine" offers an implied opposite: that which is not genuine. This use of implied linguistic "othering" does much to reinforce Genuine Jersey goods as ones that are truly representative of Jersey and ones that locals and visitors should purchase, whether to support local businesses or as a marker of remembrance.10

The inclusion of the name "Jersey" in the Genuine Jersey logo does much to convey the island as part of the product (see States of Jersey, 2010a: 33-34). The "island lure" (Baldacchino, 2010a) is inherent in the logo and therefore in the produce and products the organization supports. While Kotler (2004: 12) notes that "places are more difficult to brand than products", the Genuine Jersey brand offers a way of combining the two. The inclusion of the name of the island itself in Genuine Jersey branding offers a powerful way of simultaneously place branding and product/produce branding (Anholt, 2008; Hanna & Rowley, 2008). Compare, for example, the "New Zealand, 100%" branding exercise that was launched in 1999 (Morgan et al., 2002), which also offers a powerful linguistic message entwined in ideas of authenticity and place branding in connection with the promotion of the country as a green and clean place to visit.11

Furthermore, the Genuine Jersey logo mentions Jersey twice. First, along the top of the almost semi-circular logo the text "Product of Jersey" is written; and second, "Genuine Jersey" features as the main central wording. Linking "product", "genuine" and "Jersey" in this way reinforces the sense of the item having an authentic content, and the more goods that carry this mark then the more the idea is strengthened in comparison with goods that do not carry the mark. Moreover, not only is the term Jersey part of the overall brand, but several businesses that belong to Genuine Jersey also have the term in the title of their business or produce (Genuine Jersey, Products).

Another symbolic aspect of the logo is found in terms of its shape. While it has been designed so that the text occupies the upper part of an almost semi-circular mark, what is particularly symbolic about the shape of the lower line on the logo, which consists of a wavy or jagged line across the centre, is that this can be seen to represent part of Jersey's ragged coastline. This is another inherent aspect of the brand that allows it to stand for the island in various linguistic and non-verbal ways, and to encompass place branding simultaneously with goods' branding.

## 3.4 Activities

Genuine Jersey goods are found in many retail and public contexts. Not only do individual businesses have their own sites and mechanisms to market or sell their goods, but there are several other public contexts that do much to promote specifically Genuine Jersey products and produce. These include two shops that specialize in Genuine Jersey goods, and public events where products and produce are showcased and sold to the wider local and visiting public. Genuine Jersey has also been active in several events that reward its members, and also entice non-members to take a more active role in the organization. As noted in one of its press releases:

Nominations are being sought for three awards, namely 'Best New Genuine Jersey Product or Initiative', 'Member of the Year' and 'Genuine Jersey Retailer of the Year'. The first two awards are open to the Association's membership, which now totals almost 60 local businesses. The third award is specifically for non-member businesses only. (Genuine Jersey Products Association, 2005).

One of Genuine Jersey's founding members, La Mare Wine Estate, opened a shop in Jersey's capital, St. Helier, in 2004. The shop, which is called Maison La Mare, also sells produce belonging to other Genuine Jersey members (Genuine Jersey Products Association, 2004a). The shop is located on King Street, which is a pedestrian precinct at the heart of Jersey's retail district. Its prominent red façade along with its French name make it particularly noticeable amongst many other shops with mainly English names. Indeed, it is perhaps the shop's French name that adds a further aspect of foregrounding Jersey in terms of its geographic proximity to mainland France. Jersey makes much of its French and Norman connections, so for the visiting tourist the shop becomes a symbol for these links.

One retail site that occupies the gaze of locals and tourists alike is the shop at Jersey Airport called "The Real Flavour of Jersey" (Figure 3). The shop is a collaboration between The Jersey Pottery and La Mare Wine Estate, both members of Genuine Jersey. The shop sells some items with the Genuine Jersey logo on them, and its name further emphasizes the branding of local goods in an authentic way. Here, the "real flavour of Jersey" offers a point of reference that, like the term "Genuine Jersey", creates a hidden signifier of an "other": that is, goods that are inauthentic (i.e. non-members). Further features of the shop include its use of two Jersey flags, which are further signifiers of localness, along with several low-lying light shades that have stripes that are reminiscent of those on the French flag (Figure 3). Interestingly, another wellknown Jersey business, Jersey Pearl, has a shop at Jersey Airport just by "The Real Flavour of Jersey", although Jersey Pearl, whilst offering a brand that claims Jersey and the pearl as its own, is not a member of Genuine Jersey, presumably because of its pearls being imported to the island. However, Jersey Silver Plating is a member of Genuine Jersey, and it is based in the Jersey Pearl building.

Apart from these retail outlets, there are several regular or annual events where Genuine Jersey goods are promoted to the wider public around the island. These include the Jersey Farm and Craft Market (during July and August), St Aubin Farmers and Crafts Market and St Aubin Food Fair. Also, a special Christmas market is held a few days before Christmas (Hathaway, n.d.). While Genuine Jersey is involved with helping local businesses brand their products or produce, it also sponsors events such as the St Aubin Food Fair. This fair is the highlight of the Jersey Tourism "A Taste of Jersey in Autumn" week. In 2002, the event attracted over 10,000 people (Genuine Jersey Products Association, 2004b), and in 2003 the figure was estimated at 18 to 20,000 people (British Tourist Authority, 2003: 16).

Other market events that feature Genuine Jersey goods include the Saturday markets run by Jersey Turbot at St Catherine's; a Genuine Jersey village market as part of the Jersey Evening Post West Show; and Christmas fairs held at Jersey Pottery as part of Jersey Tourism's La Fête dé Noué (Christmas fête).

Each of the public events noted above are sites where Genuine Jersey goods are put on show to potential consumers. The events are intended for locals and tourists alike, with the majority taking place during peak tourist season. In each, Genuine Jersey is able to take preference over other traders' goods, set up stalls and promote its members' goods, or organize the events for the main purpose of promoting the products and produce of the businesses it represents. There is a distinct sense of Genuine Jersey goods at many of these events in that the brand is sometimes displayed collectively on the same stands. With these events, the government is indirectly involved in funding much of Genuine Jersey's activities, and thus gives some local businesses support at such events over other businesses who are not members of the organization (States of Jersey, 2010a: 32-33). The brand's success in this context is that it has become a wider part of island life: it is from the island, about the island and for the island. As noted by the Chief Executive Officer when asked about Genuine Jersey's biggest achievement over the past decade: "it's managed to integrate itself into the island ... it's become a part of everyday life ... it's trusted ... and ... [it] is authentic" (Garton, 2011). This statement is supported by the growth of Genuine Jersey over more than ten years: gaining more members, receiving more sponsorship, and broadening its public activities and events around the island for locals and tourists alike.

## 4. CONCLUSION

This paper has discussed Genuine Jersey as an example of a public-private organization in a small island context that exists to generate, maintain and strengthen a distinctly powerful Jersey brand. Genuine Jersey celebrates and constructs notions of authenticity, and interconnects business and government, either directly or indirectly. The discussion has shown how this organization brands Jersey produce and products within a framework of adding "island value" that either builds on consolidating recognized traditional island goods within its marketing campaign, or bringing in other goods that fulfil the requirements of being included and therefore linking them with notions of authenticity in being representative of Jersey as the place of production. Members' goods are approved to be authentically, genuinely Jersey.

As an organization, Genuine Jersey's unique place as a public-private partnership allows it to receive government support, which indirectly helps promote its members' goods. The case study has shown that Genuine Jersey is simultaneously involved in place branding in connection with Jersey, and with goods branding in connection with the produce and products of its members, some of which also include the word Jersey in their name. In doing so, the organization helps preserve goods that have long been considered traditional, and sometimes rediscovers or even invents a notion of authenticity for others within this framework. Extending the topic of the discussion even further, in the context of this paper the branding exercise has focuses entirely on produce and products (i.e. tangible items), but at a time when the global finance industry is dominating so much of the island's current income, one wonders if the label "genuine" could be applied to intangible outputs too? This is obviously beyond the scope of the current paper, but in a context where branding the island in a public-private partnership has taken on the role of place branding with an aspect of perceived authenticity inherent in membership and sponsorship criteria, questions should be asked regarding not only that criteria but also of tangible and intangible aspects of island culture that are currently not included in the Genuine Jersey brand.

While Jersey is increasingly inventing homogenous and disparate brands for itself, particularly within its three main industries – agriculture, tourism and finance – Genuine Jersey itself has crafted its own way for standing for the island. That is, it has helped to socially construct a perception of authentic goods in a marketing and branding exercise that at the same time creates an island-based value proposition with an othering of the goods of non-members (both locally produced and imported).

## ACKNOWLEDGEMENTS

I am particularly grateful to John Garton, Chief Executive Officer of Genuine Jersey (Genuine Jersey Products Association) for making time to be interviewed as part of the research process.

## REFERENCES

1. Anholt, S. (2008) 'Place Branding: Is it Marketing, or Isn't it?', Place Branding and Public Diplomacy, Vol. 4, No. 1, pp. 1-6.

2. Askegaard, S. & Kjeldgaard, D. (2007) 'Here, There, and Everywhere: Place Branding and Gastronomical Globalization in a Macromarketing Perspective', Journal of Macromarketing, Vol. 27, No. 2, pp. 138-147.

3. Baldacchino, G. (2005) 'Successful Small-Scale Manufacturing from Small Islands: Comparing Firms Benefiting from Locally Available Raw Material Input', Journal of Small Business & Entrepreneurship, Vol. 18, No. 1, pp. 21-38.

4. Baldacchino, G. (2010a) 'Editorial: The Island Lure', International Journal of Entrepreneurship & Small Business, Vol. 9, No. 4, pp. 373-377.

5. Baldacchino, G. (2010b) 'Island Brands and "The Island" as a Brand: Insights from Immigrant Entrepreneurs on Prince Edward Island', International Journal of Entrepreneurship & Small Business, Vol. 9, No. 4, pp. 378-393.

6. Baldacchino, G. (2010c) 'Islands and Beers: Toasting a Discriminatory Approach to Small Island Manufacturing', Asia Pacific Viewpoint, Vol. 51, No. 1, pp. 61-72.

7. British Tourist Authority (2003) Food and Drink in Britain, London, VisitBritain in association with Food from Britain.

8. Brychan, T. & S. Thomas. (2012) 'Cultural Events and Tourism in Jersey', Shima: The International Journal of Research into Island Cultures, Vol. 6, No. 1, pp. 114-131.

9. Cohen, E. (1988) 'Authenticity and Commoditization in Tourism', Annals of Tourism Research, Vol. 15, No. 3, pp. 371-386.

10. Corporate Edge Branding (2008) Jersey: Location Brand, http://corporateedge.com/articles/view/8

11. de Woolfson, J. (2008) 'Local Produce to get Guernsey Grown Labels', This is Guernsey, www.thisisguernsey.com/2008/ 05/16/ local-produce-to-get-guernsey-grown-labels .

12. Fornäs, J. (1995) Cultural Theory and Late Modernity, London, Sage.

13. Freire, J.R. (2009) '"Local People": A Critical Dimension for Place Brands', Journal of Brand Management, Vol. 16, No. 7, pp. 420-438.

14. Garton, J. (2010) 'The Future is Green with Genuine Jersey', in Environment Update, Vol. 3, St. Helier, States of Jersey Planning and Environment Department, p. 2.

15. Garton, J. (2011) Interview by Henry Johnson. 26 May.

16. Genuine Jersey - Application Form, www.genuinejersey.com/documents/GJPAApplicationForm2011.pdf

17. Genuine Jersey - Branding Guidelines, www.genuinejersey.com

18. Genuine Jersey - Charter, www.genuinejersey.com/documents/GJPANewCharterReleaseDec08.doc

19. Genuine Jersey - Education, www.genuinejersey.com/education.htm

20. Genuine Jersey - Our Members, www.genuinejersey.com/members.htm

21. Genuine Jersey - Products, www.genuinejersey.com/products.htm

22. Genuine Jersey - Sponsors, www.genuinejersey.com/sponsors.htm

23. Genuine Jersey Products Association (2004a) Press Release, 21 July, www.genuinejersey.com/press.htm Genuine Jersey Products Association (2004b) Press Release, 6 September, www.genuinejersey.com/press.htm

24. Genuine Jersey Products Association (2005) Press Release, 17 August, www.genuinejersey.com/press.htm

25. Grydehøj, A. (2008) 'Branding from Above: Generic Cultural Branding in Shetland and Other Islands', Island Studies Journal, Vol. 3, No. 2, pp. 175-198.

26. Grydehøj, A. (2010) 'Uninherited Heritage: Tradition and Heritage Production in Shetland, Åland and Svalbard', International Journal of Heritage Studies, Vol. 16, Nos. 1-2, pp. 77-89.

27. Hanna, S. & Rowley, J. (2008) 'An Analysis of Terminology use in Place Branding', Place Branding and Public Diplomacy, Vol. 4, No. 1, pp. 61-75.

28. Harvey, D. (1990) The Condition of Postmodernity: An Enquiry into the Origins of Cultural Change, Oxford, Blackwell.

29. Hathaway, J. (n.d.) 'Christmas Market: St. Aubin Farm & Craft Christmas Market', La Baguette, www.labaguette. org.je/ April11. php?p=2191232 .

30. Ivanovic, M. (2008) Cultural Tourism, Cape Town, Juta.

31. Jersey Tourism (2008) A Year in Review, St. Helier, Jersey Tourism.

32. Jersey Tourism (2011) Pure Jersey, St. Helier, Jersey Tourism.

33. Johnson, H. (2010) 'Branding Jèrri: Art, Image, and Identity in Contexts that Celebrate Jersey's Norman Heritage', in A. Grydehøj (ed.) Refereed Papers from the 6th International Small Island Cultures Conference, Guernsey, 23-25 June. Lismore, The Small Islands Cultures Research Initiative, pp. 13-24.

34. Kelman, I. (2007) 'Sustainable Livelihoods from Natural Heritage on Islands', Island Studies Journal, Vol. 2, No. 1, pp. 101-114.

35. Khamis, S. (2010) 'An Image Worth Bottling: The Branding of King Island Cloud Juice', International Journal of Entrepreneurship & Small Business, Vol. 9, No. 4, pp. 434-446.

36. Khamis, S. (2011) '"Lundy's Hard Work": Branding, Biodiversity and a "Unique Island Experience"', Shima, Vol. 5, No. 1, pp. 1-23.

37. Klein, N. (2000) No Logo: Taking Aim at the Brand Bullies, New York, Picador.

38. Kotler, P. (2004) 'Where is Place Branding Heading', Place Branding, Vol. 1, No. 1, pp. 12- 35.

39. Kavaratzis, M. (2007) 'From City Marketing to City Branding: An Interdisciplinary Analysis with Reference to Amsterdam, Budapest and Athens', PhD thesis, Groningen, The Netherlands, the University of Groningen.

40. Leseure, M. (2010) 'Exploitation versus Exploration in Island Economies: A Brand Diagnostic Perspective', International Journal of Entrepreneurship & Small Business, Vol. 9, No. 4, pp. 463-480.

41. Lewis, A. (2007) 'States Press on with "Banana" Logo', This is Jersey, 16 June, www.thisisjersey.com

42. Lichrou, M, O'Malley, L. & Patterson, M. (2010) 'Narratives of a Tourism Destination: Local Particularities and their Implications for Place Marketing and Branding', Place Branding and Public Diplomacy, Vol. 6, No. 2, pp. 134-144.

43. MacCannell, D. (1999) The Tourist: A New Theory of the Leisure Class. Foreword by L.R. Lippard, Berkeley CA, University of California Press.

44. Manniche, J. & Larsen, K.T. (2009) Firm-level Knowledge Dynamics of Creating Bornholm Food, EURODITE WP6 Report, Bornholm, Denmark, Centre for Regional & Tourism Research.

45. Manniche, J, Larsen, K.T. & Petersen, T. (2009) Development and Branding of `Regional Food´ of Bornholm. Final EURODITE WP5 Report, Bornholm, Denmark, Centre for Regional & Tourism Research.

46. Morgan, N., Pritchard, A. & Piggott, R. (2002) 'New Zealand, 100% Pure: The Creation of a Powerful Niche Destination Brand', Brand Management, Vol. 9, Nos. 4-5, pp. 335-354.

47. Parke, C.N. (2002) Biography: Writing Lives, New York, Routledge.

48. Pounder, P. (2010) 'Branding: A Caribbean Perspective on Rum Manufacturing Competitiveness', International Journal of Entrepreneurship and Small Business, Vol. 9, No. 4, pp. 394-406.

49. Quérée, B. (2007) '"So What Do You Think of Us Now, Chief Minister?', This is Jersey, 20 July, www.thisisjersey.com

50. Reddy, M. & Singh, G. (2010) 'Branding of Fiji's Bottled Water: Edging into Sustainable Consumption', International Journal of Entrepreneurship and Small Business, Vol. 9, No. 4, pp. 447-462.

51. Sainz, H. (2002) 'Alimentos y bebidas con denominaciones de origen y distintivos de calidad: balances y perspectivas', Distribución y Consumo, July-August, pp. 58-73, www.mercasa.es/files/multimedios/1297893230_DYC_2002_64_58_68.pdf

52. Scherrer, P., Alonso, A. & Sheridan, L. (2009) 'Expanding the Destination Image: Wine Tourism in the Canary Islands', International Journal of Tourism Research, Vol. 11, No. 5, pp. 451-463.

53. States of Jersey, Jersey Facts and Figures, www.gov.je/ Jersey/Pages/Profile.aspx .

54. States of Jersey (2007) Jersey Brand Identity Guides, www2.jersey.com/brand

55. States of Jersey (2010a) Economic Affairs Scrutiny Panel, Rural Economy Strategy 2011-2015 Review, St. Helier, States of Jersey.

56. States of Jersey (2010b) Rural Economy Strategy: Sub-panel, Rural Economy Strategy 2011- 2015, St. Helier, Economic Development Planning and Environment.

57. States of Jersey Statistics Unit (2002) Report on the 2001 Census: Jersey, St. Helier, Statistics Unit.

58. States of Jersey Statistics Unit (2010) Jersey in Figures, 2010, St. Helier, Statistics Unit.

59. States of Jersey Statistics Unit (2012) Report on the 2011 Jersey Census. St Helier, Statistics Unit. Sue's Fudge. Real Jersey Fudge, www.suesfudge.co.uk/index.htm

60. The Jersey Cow Company, www.thejerseycowcompany. com/our_story

61. The West Show Association, www.westshow.org.je/ Participation.aspx?section=takepart_local&footnote=0

62. Wang, N. (1999) 'Rethinking Authenticity in Tourism Experiences', Annals of Tourism Research, Vol. 26, No. 2, pp. 349-370.

# INDEX

## T



## U



## V